500 Years After Leonardo da Vinci Machines
Towards Innovation and Control

500 Years After Leonardo da Vinci Machines
Towards Innovation and Control

Maide Bucolo
University of Catania, Italy

Arturo Buscarino
University of Catania, Italy & CNR-IASI, Italy

Carlo Famoso
University of Catania, Italy

Luigi Fortuna
University of Catania, Italy & CNR-IASI, Italy

Salvina Gagliano
University of Catania, Italy

 World Scientific

Published by

World Scientific Publishing Co. Pte. Ltd.
5 Toh Tuck Link, Singapore 596224
USA office: 27 Warren Street, Suite 401-402, Hackensack, NJ 07601
UK office: 57 Shelton Street, Covent Garden, London WC2H 9HE

British Library Cataloguing-in-Publication Data
A catalogue record for this book is available from the British Library.

500 YEARS AFTER LEONARDO DA VINCI MACHINES
Towards Innovation and Control

Copyright © 2020 by World Scientific Publishing Co. Pte. Ltd.

All rights reserved. This book, or parts thereof, may not be reproduced in any form or by any means, electronic or mechanical, including photocopying, recording or any information storage and retrieval system now known or to be invented, without written permission from the publisher.

For photocopying of material in this volume, please pay a copying fee through the Copyright Clearance Center, Inc., 222 Rosewood Drive, Danvers, MA 01923, USA. In this case permission to photocopy is not required from the publisher.

ISBN 978-981-121-183-6
ISBN 978-981-121-245-1 (pbk)

Preface

The year 2019 signals an important date — the 500th anniversary of Leonardo da Vinci's death. He is one of the most important figures in the field of fine arts, science, architecture and engineering that existed in our entire history. Current web search results on him will show more than 180,000,000 results in 0.54sec. If you type Galileo Galilei, you will get 21,700,000 results in 0.53sec. If you type Albert Einstein, you obtain 152,000,000 results in 0.43sec. Extraordinary success is achieved by typing Bill Gates with about 216,000,000 results and an extremely high number of results only occurs for Steve Jobs (801,000,000). Michelangelo Buonarroti, another key-person in the Italian Renaissance, leads with about 7,360,000 results.

This brief search result provides us remarkable information: the permanent and worldwide popularity of da Vinci. Both as an historical personality and as a modern figure, the web indicator presents the significance of Leonardo da Vinci.

Moreover, a brief search on books on Leonardo da Vinci (only on Amazon) gives us more than 6000 volumes. Therefore, why write another volume on Leonardo da Vinci?

The idea arose in the framework of this year's celebrations of Leonardo da Vinci, together with the students of the Master Course in Automation Engineering and Control Complex System at the University of Catania. During the different courses, in fact, some peculiar aspects of da Vinci emerged. First of all, the main concept is that innovation is essential for a continuous route towards evolution in history. People who really want to innovate machines and systems should search history. Therefore, the Leonardo da Vinci inventions and projects are a suitable starting point in this aspect.

In particular, the study arising from some of the specific ideas and innovations presented by Leonardo da Vinci, led us to formulate a link with automatic control. Obviously, Leonardo himself was not able to conceive his ideas from such perspective due to the limitations in adequate theory and, especially, in the technological equipment that has been developed only in the last 100 years. Therefore, we would like to remark that the technological interest in the Leonardo da Vinci work and the educational interest for the new generation of engineers, moved us in preparing this small additional contribution to the Leonardo da Vinci literature.

Therefore, the essential aim of the book is to propose models of the Leonardo da Vinci machines, taking strictly into account the original mechanical schemes, but introducing modern low cost equipment, emphasizing the role of automatic control and that of electronic control devices, such as microcontrollers, sensors and communication devices, in order to completely automate the Leonardo da Vinci machines. The approach outlined in this book can be applied not only to other Leonardo machines not considered in our discussion, but also to other mechanical equipment not necessarily designed by Leonardo da Vinci. Moreover, it is useful to remark that the approach followed in this book can be very important also to introduce students to concepts typical of automation and for assisting in learning, keeping in mind the practical applications of advanced automation principles.

Through the presentation of a novel and a wide point of view (even if incomplete!) on the Leonardo da Vinci studies, we will remark on the modernity of Leonardo da Vinci with regards to his approach towards science. It has been recently remarked from several authors [Kemp (2007); Taddei (2007)] that the Leonardo da Vinci science is very close to the current Systems Science theory. Indeed, in the Master Course on Automation Engineering and Control of Complex Systems at the University of Catania, the control of complex systems leads us to rediscover and review several key knowledge points of the engineering principles, in particular that remarked by R. Buckminster Fuller: *Act locally, think globally* [Fuller (2001)]. Reviewing the paradigm of the Leonardo da Vinci studies and the cornerstone book on complexity by Klaus Mainzer [Mainzer (2007)], we can discover a lot of common points after 500 years!

Is Leonardo the father of Complex Systems Engineering? This is another aspect that will be deeply discussed in the book.

Summarizing, this book will consider some items of the Leonardo da Vinci work and discuss critically on the impact that, up to now, his ideas have had. From an educational point of view, models of Leonardo da Vinci machines have been taken into account in order to prove how innovation and history can be successfully linked. Indeed beyond the scientific and cultural aims, the role of amusement is also emphasized, that definitely is an essential ingredient in order to improve the appeal of educational activities.

The book is organized into seven chapters. They follow the genesis of the book. Chapter 1 includes the first model of the machine that has been realized: the Leonardo da Vinci flying machine with amusing experiments. In Chapter 2 the dualism between the modern paradigm of complexity science and Leonardo's studies will be discussed. The peculiar aspects that are fundamental to understand Leonardo's machines and their design as related to the organization of eyes and mind of da Vinci are discussed in Chapter 3. A discussion on the technical design of Leonardo as a powerful tool to invent and dynamically represent projects is found in Chapter 4. Chapter 5 outlines the projects of Leonardo showing the great vision and foresight of Leonardo da Vinci as civil engineer, architect and urban planner. In Chapter 6, the technical scenario at the Leonardo age that established the framework of Leonardo inventions is evaluated. The key Chapter 7 reports the selected machines' models that have been realized and innovated. The task has been identified and further machines inspired by the Leonardo models are considered. The conclusive remarks reported in the last Chapter 8 stress the importance of this work both from a scientific point of view as well as the straightforward impact from the educational point of view. A list of related projects is included and initiatives in progress will be presented. Finally the book has two appendices: Appendix A reports a concise general biography of Leonardo da Vinci, in Appendix B, the technical details on the realized equipment included in Chapter 7 are discussed.

M. Bucolo A. Buscarino C. Famoso L. Fortuna S. Gagliano

Contents

Preface v

1. Leonardo's Flying Machine: model and pleasant details 1

2. Thinking in Complexity and Complex Systems Engineering with Leonardo da Vinci 5
   - 2.1 Cosmology 8
   - 2.2 Quantum physics 9
   - 2.3 Hydrodynamics 10
   - 2.4 Meteorology 11
   - 2.5 Geology 13
   - 2.6 Chemistry and material science 14
   - 2.7 Biology 15
   - 2.8 Sociology and economy 16
   - 2.9 Neuroscience 16
   - 2.10 Computer science 17
   - 2.11 Some conclusive remarks 17

3. The Eye and the Mind of Leonardo da Vinci: some remarks 21
   - 3.1 Main concepts 21
   - 3.2 The vision in the Leonardo da Vinci brain 23

4. Leonardo da Vinci and the Technical Design 27
   - 4.1 Some notes 27

5. Leonardo da Vinci: civil engineer, architect and urban planner 31
    5.1 Remarks and discussion 31
6. Leonardo da Vinci and the Engineering Scenario at His Time 39
    6.1 Some notes 39
7. A Selected Number of Machines: innovation and control 43
    7.1 The Leonardo da Vinci machines 43
    7.2 Drop propagation 44
    7.3 Mechanical clock 56
    7.4 The Lever Crane 61
    7.5 Rolling ball timer 61
    7.6 Mechanical drum 63
    7.7 Flying Machine 66
    7.8 Self-supporting bridge 66
    7.9 Transformation of motion 68
    7.10 Final considerations 71
8. Epilogue 73

Appendix A Appendix I - Biographic Notes on Leonardo da Vinci 77

Appendix B Appendix II - Circuits and Microcontroller Codes 79
    B.1 Bluetooth module 79
    B.2 Drop propagation 80
    B.3 Mechanical clock 82
    B.4 Lever crane 84
    B.5 Mechanical drum 88
    B.6 Self-supporting bridge 89
    B.7 Transformation of motion 92

Bibliography 95

Chapter 1

Leonardo's Flying Machine: model and pleasant details

Perhaps, the most impressive machine invented by Leonardo is the flying machine. This machine has been linked to the main dream of flying and in particular to soar the sky like the birds.

The Leonardo da Vinci concept of flight assumed a mechanical meaning rather that a heroic or a divine gift. In Fig. 1.1, two pictures of the Leonardo flying machine realized during our voyage in the Leonardo da Vinci world are included. The Leonardo flying machine can be considered as the introduction of the machinery in the history of flying.

The main studies of da Vinci on flight are included in the *Codex on the flight of the birds*. It is a short codex dated 1505 containing 18 folios. The codex is now located in the Museum of the Biblioteca Reale in Turin. In this codex, Leonardo investigates and explains some particular physical aspects related to the mechanics of flight. He discusses the gravity, the density of air, the concept of force balance and also oscillations. Leonardo understood that a fundamental source of energy is in the air in fluid dynamics.

From our perspective, Leonardo da Vinci's main project was to invent an equipment able to harvest energy from the environment. During his time,

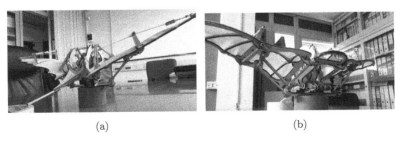

(a)   (b)

Fig. 1.1 The Leonardo da Vinci flying machine model realized by using wooden elements and ropes.

in fact, converting and storing energy was a main problem, due to the lack of suitable machines. Therefore, taking into account that the unique type of engine existing during the Renaissance age was the spring-based engine, and considering its great limitations, Leonardo's idea was to use mechanical equipment that were able to convert the air currents to obtain coordinated motion. Indeed, Leonardo himself was convinced that his invention could never be powered by a man that cannot create enough power for take-off of the machine.

It is in Folio 7 of the *Codex on the flight of the birds* that Leonardo starts to describe the flying machine, making a comparison with his notes on the birds' flight and a possible artificial wing, discussing also about the materials that could be used in order to realize such a type of machine. Indeed, when presenting the model of Leonardo flying machine during the course, we were led to consider that the model of the Leonardo machine marked a milestone in the scientific thought and therefore we decided that the model deserved to fly!

During a semester of lessons, laboratory activities, and experiments we concluded that free experiments are fruitful both for the teachers and students. First of all, innovation must be combined with sustainability and therefore each action must be accomplished under the previous assumption.

In order to give flight to the Leonardo machine, more solutions were considered. The first one was to load the Leonardo machine on a drone. This solution was immediately discarded since due to the weight of the machine, about 500g, a powerful drone had to be used. Moreover, we wanted to respect the Leonardo principle to use sustainable energy.

Therefore, the choice was to use a kite for the first flight of our Leonardo machine. We used a cellular kite as the model in Fig. 1.1, equipped with a small camera to record a video. The Leonardo machine, indeed, recorded a video showing the coastal area of the Catania Playa beach. A series of frames are observed in Fig. 1.2.

In a following experiment, the flight machine of Leonardo took off by means of three low-weight balloons filled with helium, each of about 80cm of diameter. A series of pictures of the experiment and the views of Catania taken from the machine are reported in Fig. 1.3. The balloons allowed the Leonardo machine to reach about 150 meters from the ground.

A brief remark: during the flight three or more real birds approached the Leonardo machine. Were they really similar? Were the birds confusing the machine with a predator or trying to follow it? The flight of the Leonardo machine was a start!

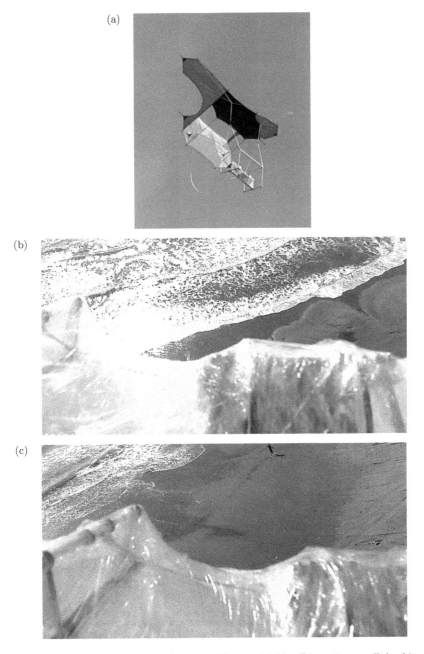

Fig. 1.2 The Leonardo da Vinci flying machine model lift off by using a cellular kite: (a) flying machine, (b)–(c) views of Catania Playa beach taken from the flying machine.

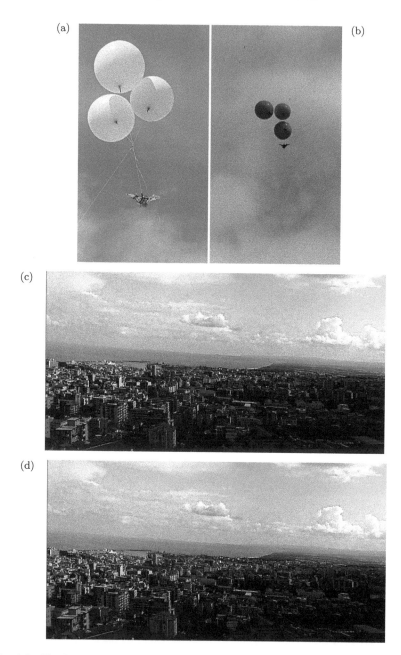

Fig. 1.3 The Leonardo da Vinci flying machine model lift off by using three helium balloons: (a) flying machine, (b) a series of birds approaching the flying machine, (c)–(d) views of Catania taken from the flying machine.

Chapter 2

Thinking in Complexity and Complex Systems Engineering with Leonardo da Vinci

The first aim of the book *Thinking in Complexity* by Klaus Mainzer, first published in 1994, is clarified by the statement "The new science of complexity would characterize the scientific development of the 21th century" [Mainzer (2007)]. After a quarter of a century, the sentence proved to be essentially true. Many courses on "complexity" are present today in the universities worldwide. Moreover, the concept has been widely included in engineering where the emergence of complex phenomena arising in science, is essentially linked to nonlinearity and to interactions between simple systems [Buscarino *et al.* (2017)].

This is remarkably evident in the Internet age, especially in the concept of Internet-of-Things (IoT) and in the design of networked machines where the behavior of the single unit does not characterize the emergent behavior of the whole system. Indeed, the architect Richard Buckminster Fuller established a paradigm in civil engineering proposing the sentence "Act locally, think globally" [Fuller (2001)].

The book *Thinking in Complexity* outlined the palimpsest of complexity and highlighted the efforts that led, after the conception of the system theory, toward this new type of science emerging in the scientific world. In the book, it is stressed that the general methods of nonlinear complex systems must be developed in cooperation with more and more sciences, in order both to stimulate the mind of the scientist in conceiving the general aspects of more diverse problems and to include in the background of research the possibility of predicting, modeling, designing new systems with performances of extreme importance today: the sustainability of the planet and the interaction among the people.

In the last years, several research centers dedicated to the study of complexity have been created in the world, we cite only some of them:

- Complex Systems Society
- Center for Complexity Science, Imperial College London (UK)
- Center for Theoretical Physics, CNRS, Marseilles (France)
- Complex Networks Group, University of São Paulo (Brazil)
- Complex Systems and Networks Lab (Cosnet), U Zaragoza (Spain)
- Center for Complex Network Research (Barabasi Lab), Northeastern U (Boston, MA)
- Center for Chaos and Complex Network, (Hong Kong (PRC)
- Institute for Scientific Interchange (ISI Foundation), Turin (Italy)
- Max Planck Institute for Dynamics and Self-Organization, Göttingen (Germany)
- Max Planck Institute for the Physics of Complex Systems, Dresden (Germany)
- New England Complex Systems Institute (NECSI) (Boston, MA)
- Santa Fe Institute (Santa Fe, NM)
- Stanford Complexity Group, Stanford U. (Palo Alto, CA)
- a permanent Master Course in Automation Engineering and Control of Complex Systems is taught at the University of Catania (Italy).

A question arises: was Leonardo da Vinci the first person really thinking of complexity?

Elements leading to something similar to the concept of complexity can be found in the ideas of the ancient pre-Socratic philosophers Thales of Miletus (625 – 545 BC) and Anaximander (610 – 545) and so on, the thinking in complexity emerged from discussions and notes. Moreover, the idea of complexity took a concrete role in the design of complex engineered systems only thanks to Leonardo da Vinci. Leonardo da Vinci inadvertently approached his study thinking in complexity, and his projects, designs and inventions have been conceived in the framework of complex systems, even if, at that time, the term was not known. Naturally and without precise knowledge, Leonardo opened the route of complexity and deeply marked it.

Under this new perspective, we now want to analyze the work of Leonardo bearing in mind the outstanding table reported in the book *Thinking in Complexity*, which clearly explains the palimpsest of complexity, as in Fig. 2.1. We will show that a parallel palimpsest corresponding to the Leonardo investigations emerges.

Table Interdisciplinary applications of nonlinear complex systems

DISCIPLINE	SYSTEM	ELEMENTS	DYNAMICS	ORDER PARAMETER
cosmology	universe	matter	cosmic dynamics	cosmic pattern formation (e.g., galactic structures)
quantum physics	quantum systems (e.g., laser)	atoms photons	quantum dynamics	quantum pattern formation (e.g., optical waves)
hydrodynamics	fluids	molecules	fluid dynamics	form of fluids
meteorology	weather climate	molecules	meteorological dynamics	pattern formation (e.g., clouds hurricanes)
geology	lava	molecules	geological dynamics	pattern formation (e.g., segmentation)
chemistry	molecular systems	molecules	chemical reaction chemical dynamics	chemical pattern formation (e.g., dissipative structures)
materials science	smart materials nano systems	macromolecules	macromolecular dynamics	macromolecular pattern formation (e.g., nano forms)
biology	genetic systems organisms populations	biomolecules cells organisms	genetic reaction organic growth evolutionary dynamics	genetic pattern formation organic pattern formation pattern formation of species
economics	economic systems	economic agents (e.g., consumer producer)	economic interaction (e.g., mechanisms of markets)	economic pattern formation (e.g., supply and demand)
sociology	societies	individuals, institutions etc.	social interaction historical dynamics	social pattern formation
neurology psychology	brain	neurons	neural rules learning algorithms information dynamics	neural pattern formation pattern recognition
computer science	cellular automata neural networks global networking (e.g., Internet, ubiquitous computing)	cellular processors	computational rules evolutionary algorithms learning algorithms information dynamics	pattern formation of computational networks

Fig. 2.1 Palimpsest of Thinking in Complexity [Mainzer (2007)].

2.1 Cosmology

Leonardo da Vinci discovered the phenomenon of Earthshine that can be defined as the reflection of the sunlight from the Earth towards the Moon. It occurs when there are conditions under which the light from the Sun is reflected by the Earth surface to the Moon and then goes back to our eyes. To discover this phenomenon, 500 years ago, a great imagination was required. We remark that the copernican sun-centered system was presented in 1543!

The Leonardo da Vinci discovery was the result of merging together more strictly related studies: optics, geometry and observation. His conclusions are reported by using drawings, as shown in Fig. 2.2, that are included in Leonardo's *Codex Leicester* where he commented on the page "Of the Moon: no solid body is lighter than air". Leonardo da Vinci concluded that the Moon has an atmosphere and oceans. Moreover, Leonardo da Vinci made a mistake! Moon does not contains oceans as it was discovered by the Apollo 11 astronauts just 50 years ago.

Fig. 2.2 Sketch of Earthshine performed by Leonardo da Vinci (Codex Leicester).

The answer given to the mystery of Earthshine by Leonardo includes the conceiving of the system, the materials and the dynamics and patterns that are the main guidelines to include in this study of da Vinci as a complete example of thinking in complexity with respect to the cosmology discipline.

2.2 Quantum physics

Even if in the Leonardo era, classical physics had not yet been formalized and Newtonian physics arose some centuries later, important algebraic principles that will be at the core of quantum mechanics had been explained by Leonardo da Vinci. As reported by Hermann Weyl in [Weyl (1950)], Leonardo da Vinci just introduced the Theory of Groups. The group expresses the isotropy or homogeneity of the space: it consists of all-to-one isomorphic correspondences of the space on itself, it describes the transformations which leave invariant all objective relations between the points of the space expressed geometrically. It is interesting to note that the theory of groups assumes a role in fine arts: the group-theoretic nature of the Egyptian art can be considered as a shining example [Weyl (1950)]. Leonardo da Vinci worked on formulating the formal principles of art, and therefore to give a general and systematic approach in order to characterize the various possible symmetries in a building.

Indeed, quantum mechanics widely uses group theory in quantum physics. The relationship between quantum physics and Leonardo turbulence studies have been recently remarked in [Barenghi et al. (2014)], where we see that the quantum turbulence appearing in quantum fluids is clearly designed by Leonardo da Vinci.

Indeed, da Vinci established that observation and an accurate drawing can give reasonable ideas of the mechanisms that drive the phenomenon. The relationship between Leonardo and Quantum physics also emerges from the universal concept explained by Leonardo da Vinci himself as "Developing a complete mind" [Lupacchini and Angelini (2014)]: *"Study the science of art. Study the art of science. Develop your senses, especially learn how to see. Realize that everything connects to everything else"*. Moreover in [Tonomura (1998)] it is further remarked that Leonardo da Vinci was a precursor of quantum physics.

Why cannot we explore quantum physics through art and getting people to think? This was a project of Ben Still and Lyndall Phelps who realized an artistic contemporary installation focused on physical science from the microscales to macroscales: the fellowship between art and science initiated by Leonardo stands still strong.

2.3 Hydrodynamics

The studies of Leonardo da Vinci about hydrodynamics focused on more topics, but, of course, the subject of turbulence received the most prominent attention. The method used by Leonardo in order to investigate a subject was set out on observation and drawing. In conceiving the particular dynamics of water turbulence, Leonardo performed several accurate drawings of the phenomenon. Leonardo paid great attention during the years 1508–1513 to the study of turbulence and the corresponding patterns produced by water flowing from a canal and entering a body of water from above its surface [Monaghan and Kajtar (2014)].

Fig. 2.3 Leonardo study on water vortex and hydrodynamics (Codex Windsor).

Turbulence is one of the oldest unsolved problems in physics. In order to say that the problem is solved, we have to prove that we can predict the fluid motion based on their equations and on the initial conditions. In principle, it should be completely solved, however the transition between laminar and turbulent regime is until now not truly solved. Even if sophisticated instrumentation is now available in laboratories in order to visualize the phenomena, and supercomputers are available to perform numerical simulations, the drawings produced by Leonardo da Vinci still present turbulence in a more realistic way. Indeed Leonardo da Vinci understood in depth the complexity of the problem and approached it with his method: observation and drawing as that reported in Fig. 2.3.

Only recently a problem, known as hydraulic jump widely investigated by Leonardo da Vinci, has been solved [Bhagat *et al.* (2018)]. A hydraulic jump does occur in water flowing into a sink and Leonardo deeply analyzed it, as shown by the highly precise drawing in Fig. 2.4.

Fig. 2.4 Leonardo study on hydraulic jumps (Codex Windsor).

Indeed the interest in fluid dynamics of turbulence and violent waves was related to the aim of Leonardo da Vinci of saving energy and during his time, the more useful energy was in the water.

2.4 Meteorology

In Fig. 2.5, an example of the Leonardo da Vinci clouds are reported. In the current literature, Leonardo da Vinci, whose studies on clouds had been viewed as the result of a multidisciplinary genius, is considered as the first modern meteorologist. Moreover, we focus our attention on the relationship between cloud shapes and water turbulence. The Leonardo da Vinci mind worked as that of a complex system scientist. Looking at the correlations among various behaviors of water and clouds, Leonardo da Vinci tried with his drawings to establish a general model of the phenomenon. He understood that both water turbulence and clouds have common points and that both are fundamentally dynamical complex systems. However, during his time the necessary mathematics to describe the correct model of the phenomenon was not available. Again Leonardo is able to propose an effective model by using his peculiar instrument: drawings.

Fig. 2.5 Leonardo study on clouds (Codex Windsor).

On the basis of a mathematical model and the related boundary conditions, modern computer programs allow to integrate the equations, to make a prediction and, thanks to the computer graphics tools, to have a representation of the phenomenon. The mind of Leonardo worked in a similar way and he was able to provide studies both of clouds and the water turbulence. His consideration is straightforward: "All our knowledge has its origin in our perception". Moreover, Leonardo da Vinci clouds are related to what occurs before and during a storm and deeply describe the phenomenon of catastrophes related to the clouds.

It is important to note that the pictures of clouds presented by Leonardo are not only the expression of some phenomena, they are a dynamical description of energy release. This reinforces our opinion that Leonardo was mainly interested in the phenomena of energy transformation or transfer, something which became dominant in his life.

2.5 Geology

In the studies of Leonardo, an appealing but unusual topic at that time, was geology. The Leonardo studies are often connected with one another, a further key point to assert Leonardo as a complexity scientist. Leonardo in our opinion was not interested in a subject just for curiosity. His researches were related to each other and geology is related to the Leonardo study on hydraulics and to his hydraulic projects, such the diversion of the Arno River, as seen in the drawing in Fig. 2.6. Moreover, as an architect and city designer, Leonardo da Vinci needed to understand geology.

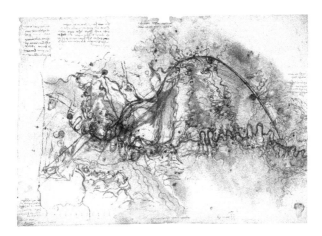

Fig. 2.6 Leonardo study on the diversion of the Arno river (Codex Windsor).

Again the drawing plays a fundamental role, also with regards to Leonardo as a geologist. In Fig. 2.7, is Leonardo da Vinci's rock ravine. He studied the erosion and the dynamics that led to the actual scenario of mountains and caves. He studied also fossils and shells. By using his observations, Leonardo established the dynamic evolution of the Earth. Moreover, his studies emphasize the power of water and how the force of water makes great natural sculptures. For this, nature needs a lot of time. From this observation, he argued the age of the planet.

Fig. 2.7 Leonardo study on cliffs and ravines (Codex Windsor).

The history of the planet is signed by fossils. He discovered in the fossils the marked sensors of the geological ages.

The idea about the Earth by Leonardo da Vinci was a pulsing model with the surface going up and down, a dynamical model also justified by geometrical considerations that explains the presence of sea shells on the top of the mountains.

Also regarding the geological research topic, the Leonardo da Vinci approach is related to modern complexity approach: again act locally, think globally.

2.6 Chemistry and material science

As a painter, Leonardo focused several studies and experiments on colors. In particular, in the case of glazes he was able to realize thin layers of $1 \div 2$ μm. Referring to this particular topic, Leonardo da Vinci can be considered a modern scientist of the surface. This study was recently performed by the team of laboratories of the Centre de recherche et de restauration des musées de France by using x-ray fluorescent spectrometry [De Viguerie et al. (2010)].

It is also reported that Leonardo was able to synthesize a chemical compound similar to acetone and even bakelite cement can be obtained by following his notes.

2.7 Biology

The Leonardo contributions in biology are wide and comprehensive. They regard several fields including botany, zoology and anatomy. The approach adopted is not merely descriptive. The Leonardo studies are mainly focused on relating some behaviors in order to understand physiology. The main difference in his approach in studying this field, with respect to other people, is that he searches the processes continuously that characterize the living systems.

For example, in botany, he tried to relate the morphology with the physiology of the plants. He answered himself the question on how the plants obtain energy in order to activate their growth. Leonardo noticed what is the external stimulus that the environment plays on the plants. These are outstanding results that allowed to propose, at that time, quite new paradigms of knowledge. Leonardo again can be considered the first modern botanist and the first ecologist.

The approach used in zoology was the same. We remark here that the study of Leonardo da Vinci had been strongly consolidated to characterize the dynamical movements of animal behavior, in particular, in understanding the physiology of movement in order to translate it in biologically inspired machines.

Moreover, the study of Leonardo da Vinci was not restricted only to this aim. His studies regarded also the comparison between human and animals skeletons and muscles. He achieved also important observations on animal perception, discovering sensory behavior in the animals. It is particularly surprising to note his observations on the animals' social behavior. In our opinion, the study of crabs, that was done during his stay in Venice, anticipated the approach with modern multi-legged robots. He was actually fascinated by the mechanics of movement of crabs. Moreover, a lot of animal drawings from Leonardo proves this strong interest in zoology.

The studies of anatomy performed by Leonardo are also surprising. The accurate drawings of the various human organs are universally accepted as outstanding. The interest of Leonardo da Vinci in the *human apparatus* has been strongly related to his major skills in painting. Moreover, the approach of Leonardo da Vinci in anatomy studies has been the same of other fields: pursuit to understand. Therefore, again the anatomy studies are strongly related to the physiology.

2.8 Sociology and economy

Even if the science of sociology was born two centuries after Leonardo's death, thanks to the studies of Auguste Comte, Montesquieu and Jean-Jacques Rousseau, in the XVIII century, it is conjectured that in the mind of Leonardo da Vinci some concepts related to this discipline already existed. In fact, Leonardo could not have been a town planner without having some precise references in disciplines like sociology within a political context.

In conceiving the modern idea of city and in the introduction of regulatory plan of Florence (Codex Windsor 12681) both with its dodecagon perimeter including the regular network of orthogonal primary roads, and with the Arno river rectified route, he gave a new paradigm of the city, showing that more disciplines were included in his urban planning model. Economy, transportation science, living disciplines and, first of all, the connection of beauty with concreteness, led to establish some new design concepts in the city planning, perhaps unifying other people's studies with the historical and the economical aspects. This is dominant in his universal concept of city.

A complete idea on a new economy based on efficient transportation systems, with a rational conception of the city, was also a key point in the perspective mind of Leonardo da Vinci as an engineer. More inventions emerged from the efforts of Leonardo in building new cities and efficient transportation systems. Inside the mind of Leonardo, these aspects are all related: in building a city, it is necessary to have new big and efficient mechanical equipment, to create reliable river transportation, new hydraulic machines and transportation infrastructures.

These are the "springs" in Leonardo mind in order to adopt a complex systems approach in systems engineering.

2.9 Neuroscience

Leonardo's studies in the area of neuroscience are pioneering. He discovered in the brain the frontal sinus and the meningeal vessels. By injecting hot wax into the brain of an ox he discovered ventricles. He developed an original mechanist model of sensor physiology. His research was finalized by discovering how the brain processes vision and other sensory inputs and how brain integrates these information [Pevsner (2002)].

2.10 Computer science

Leonardo da Vinci's inventions include also some mechanisms that can be adopted for calculating purposes. In the Madrid Codex I (folio 36v) drawings of several mechanical equipments to develop a mechanical calculating machine are reported, as in Fig. 2.8.

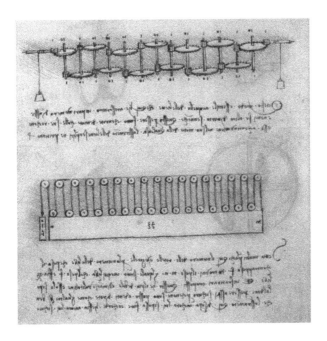

Fig. 2.8 The Leonardo study on mechanical calculating devices (Codex Madrid I).

2.11 Some conclusive remarks

The previous notes on the Leonardo studies follow the same outline as in the complexity table shown in Fig. 2.1. The Leonardo approach is the same as used in modern complexity science and complex engineering. In order to highlight the impressive role of Leonardo da Vinci in the actual complexity field, we take into account the Codex Atlanticus (1119 pages) that is considered the most exhaustive collection of notes and drawings and designs of Leonardo da Vinci. At the web address codex-atlanticus.it/#/overview we can analyze page by page the interdisciplinary nature of the study of Leonardo da Vinci. It is possible to select among the topics: Geometry and Algebra, Physics and Natural Sciences, Tool and Machines, Architectures and Applied Arts, Human Sciences. We can have a complete view on how many pages go into each subject and the colored page patterns shown in Figs. 2.9 and 2.10, give us a clear indication of the completeness of Leonardo da Vinci studies.

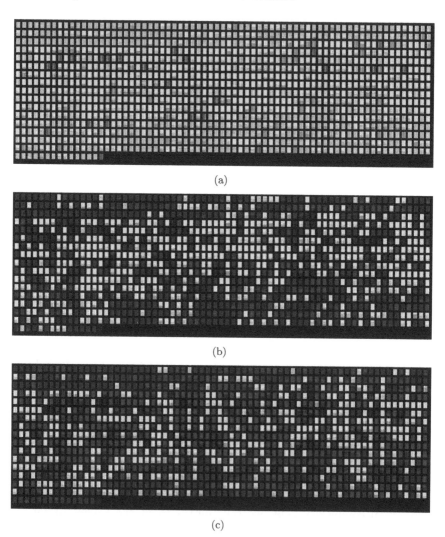

Fig. 2.9 Leonardo da Vinci Codex Atlanticus: (a) complete scenario of the 1118 pages, occurrences of subject (b) Geometry and Algebra (1141 times in 515 pages), (c) Physics and Natural Sciences (1004 times in 394 pages).

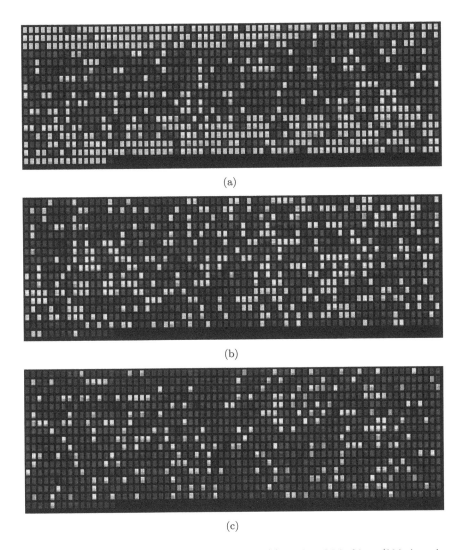

Fig. 2.10 Leonardo da Vinci Codex Atlanticus: (a) Tool and Machines (904 times in 504 pages), (b) Architectures and Applied Arts (496 times in 361 pages), (c) Human Sciences (429 times in 236 pages).

The numerical results and patterns in Figs. 2.9 and 2.10 give us a quantitative point of view on the interlacing Leonardo studies that is the fundamental aspect of the emergent science of complexity.

Chapter 3

The Eye and the Mind of Leonardo da Vinci: some remarks

3.1 Main concepts

Andrea del Verrocchio (1435 Florence–1488 Venice), Florentine sculptor and painter, was Leonardo da Vinci's teacher. He said about Leonardo's Eye: "It appears that his eyes perceive something that eyes of others could not see".

About Leonardo's Eye, a lot has been said and written [Ackerman (1978)]. Moreover, the sentence of Verrocchio globally explains the world and life of Leonardo da Vinci. The emergent traits of his human observations are that particular skills are enhanced with genetic origin and, if elicited, allow people to achieve excellence in particular fields. The brain plasticity, related to the synaptic plasticity, consists in the possibility of modifying the strength of the synaptic connections in the brain. Therefore, our brain is continuously evolving, adapting itself to external events, including experience. This is obtained by varying synaptic connections, either inhibiting or enforcing some of them. The mechanisms of the neural plasticity are fundamental in the case of neural disease or neural injuries, in these cases the self-organization of the brain (reinforced by the plasticity) allows emerging surprising properties of the brain in order to strengthen some brain capabilities to compensate for other brain deficits [Maffei and Fiorentini (2008)].

In his book, the neuroscientist Lamberto Maffei makes an important contribution in a chapter dedicated to the eye's theme. The title of the chapter is "The frog's eye and the Galileo Galilei eye".

Neuroscience recently proved that the evolution process, in millions of years, acts on the eye. In the eye of animals, like that of the frog, more tasks are concentrated. The eye is a sensory organ, moreover in the eye of the frog, processing activities on the perceived image are performed. This allows living creatures at the bottom of the evolutionary scale to respond to the perceived situations with immediate actions. The eye of humans, and of course the eye of Galileo Galilei, is limited with reference to this point of view, in fact the real processing of information is performed by the brain in the visual cortex. Moreover in human beings, a positive feedback does exist. In some sense, it is our brain that told us what to see. It is mainly the visual cortex that tells us what to see and how to see with reference to the details of the image we are sensing. It is the visual cortex that is the primary source of image information and therefore originates the environmental–man reaction. The interpretation of the images is hence mediated by human knowledge. It is the brain that addresses (controls) the eye in order to enhance some sensory features rather than others.

The observations made by Galileo Galilei on Moon features and his drawings explain clearly this concept. Using the telescope that he invented, Galileo was able to reproduce particular features of the Moon, like craters and mountains. The drawings of Galileo Galilei are very similar to the images that today are transmitted by satellites and lunar probes. The knowledge of Galileo about shading and the geometric approach to drawing, lead to the Galileo Moon being realistic. It is a real fact that cultural mediation plays an important role in seeing and in representing what is seen.

The Galileo Galilei (1564–1642) moon drawings (Nov–Dec 1609), as in Fig. 3.1, are located at the Biblioteca Centrale di Firenze and represent the first realistic view of the Moon in history.

Moreover, the astronomer Thomas Harriot (1560–1621) in the same period 1609–1611 also drew the Moon. Even if the drawings are very similar, the Harriot images, as in Fig. 3.2 are much simpler and reveal a limited capability of eye processing with respect to the Galileo representation.

The previous consideration indicate the significance of Andrea del Verrocchio's comments about the Leonardo da Vinci eye. The actual knowledge of neuroscientists, and in particular, the Maffei observations related to the role of cultural mediation in sight, lead us to argue that the optic knowledge, the perspective studies and the knowledge of the drawing techniques combined with the genetic predisposition, allow scientific confirmation of the perspective of Leonardo da Vinci.

The Eye and the Mind of Leonardo da Vinci: some remarks 23

Fig. 3.1 Galileo Galilei drawings of the Moon surface.

3.2 The vision in the Leonardo da Vinci brain

The Treatise on Painting of Leonardo da Vinci [Vasari *et al.* (1890)] is the first on neuroesthetics in history, as Semir Zeki confirmed [Zeki and Nash (1999)]. The Treatise was recommended to be published by Francesco Melzi in 1540. It consists of 935 observations on paintings, organized in notes.

Some important notes on colors examine the topic of color contrast. Leonardo da Vinci discovered the principle of complementarity, that was only recently formalized by Svaetichin and Jonasson [Svaetichin and Jonasson (1956)]. In accordance with this principle, the cells in the retina excited by the red color are inhibited by the green color, the cells excited by the yellow color are inhibited by the blue color, the cells excited by the white

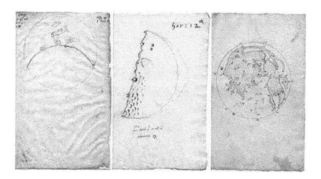

Fig. 3.2 Thomas Harriot drawings of the moon surface.

are inhibited by the black (and vice versa). This principle recently has been explained in terms of neuronal behavior, this means that the neuron cells of the visual cortex can modify their reaction in accordance with the background color against which is presented the color for which they are selective [Zeki and Nash (1999)]. Even if these principles have been recently formalized, they had been intuitively discovered by Leonardo da Vinci.

More insights can be found both in the book of Leonardo da Vinci and in the actual reference book on *Art and Brain* by Semir Zeki: "The eye, the window of the soul, is the chief means whereby the understanding can most fully and abundantly appreciate the infinite works of nature". [Zeki and Nash (1999)]. This sentence strongly emphasizes the role of the eye in Leonardo da Vinci art philosophy. In fact Leonardo da Vinci, strongly motivated by his main interest as an artist, studied into detail both the optic principles and the anatomy and physiology of the eye. In the Codex Atlatincus he annotates "The eye, of which experience shows us so clearly the office, has been defined by many time by an infinite number of authors of a given way and find that it is completely different." (Atlantic Codex fol 361 V).

Many studies and citations of Leonardo, collected in more notes, proved that he anticipated the Galileo Galilei eye concept. "Make glasses to anticipate big Moon", he wrote (Atlantic Codex, fol. 190 1a). The main notes of Leonardo da Vinci are collected in ten codes:

- Codex Arundel — 283 pages (1478–1518) — Studies on physics, mechanics, optics, geometry and architecture;
- Codex Atlanticus — 1119 pages (1478–1518) — Studies on astronomy, optics, mathematics, botany, zoology;

- Codex Trivulzianus — 55 pages (1478–1493) — Studies on architecture and literature;
- Codex on the flight of birds — 18 pages (1505) — Studies on the analysis of flight, wings shape, air resistance;
- Codex Ashburnham (also known as French Manuscripts) — 283 pages (1489–1492) — Studies on painting and physics;
- Codices of the French Institute — 12 volumes, 964 pages (1478–1518) — Studies on geometry, hydraulics and optics;
- Codices Forster — 3 volumes, 94 pages (1478—1518) — Studies on hydraulic machines and geometry;
- Codex Leicester — 36 pages (1506–1510) — Studies on astronomy and hydraulics;
- Codex Windsor — 600 pages — Studies on astronomy, drawings, caricatures;
- Codex Madrid — Madrid I, 384 pages, Madrid II, 316 pages (1490–1505) — Studies on geometry, statics and mechanics.

In each of these codes, the eye subject is mentioned. This is to confirm the importance of the eye for Leonardo da Vinci. From the continuous rearrangement of the Leonardo da Vinci codes, we observe the continuous deepening of the importance of optics and geometry incorporated into the drawing and the painting.

It is well known today the laws of the visual cortex in the human brain. Why do we see? We see to acquire the knowledge of our world [Zeki and Nash (1999)]. This remark is particularly appropriate from the Leonardo da Vinci point of view. We can add a further sentence: Art is the search for the essence of the universe. The Art and the drawing are creative processes that are an extension of the visual brain behaviors.

Chapter 4

Leonardo da Vinci and the Technical Design

4.1 Some notes

Design must be considered a fundamental mechanism, strictly related to a cognitive process. The development and evolution of the human brain are strictly related to the evolution of the hand [Young (2012)]. The hand realizes the processes that materialize the ideas. Moreover, the hand initiates some processes that can create new ideas. This is the reason why Leonardo da Vinci made a lot of effort in drawing and studying hands, in accordance with the actual theories of neuroscientists. One of the wonderful drawings among the hand studies made by Leonardo can be seen in Fig. 4.1.

Fig. 4.1 Leonardo study of two clasped hands (Codex Windsor).

Hand and Brain, Hand and Drawings. These have been the paradigms of the main conception that characterized the essential meaning of Leonardo da Vinci. "Drawing is the tool that allows the eye to reach the maximum clarity and bring the mind maximum precision" claims Henry Matisse [Finsen (2005)].

Thinking and drawing share many items. Addressing and confining by ordering the continuous flow of ideas that had not yet taken a new form is the ultimate reason for drawing. In our brain the flows of thinking, sensorial data, memory traces, projections and unconscious instincts lead to complex patterns of connectivity between neurons. As neuroscience has proved, drawing helps the organization of this process.

Both the capabilities and the development of the hands and of our brain are increased together, thanks to drawing. Drawing is an immediate function to be performed. Drawing is the immediate artificial language that our mind perceives.

"We were never born to read". This is the first sentence in the neuroscientist Maryanne Wolf's book [Wolf and Stoodley (2008)]. Indeed she wrote also: "The genes associated with dyslexia have survived vigorously". The behavioral neurologist Normand Geschwind often remarks in his studies that the diversity of skills and deficits we inherit, contributes to the construction of a society capable of satisfying everyone's various needs. Dyslexia is an example of the human diversity, it is of great importance in the process of improving the development of human culture. Indeed more neuroscientists have proved dyslexia in Leonardo da Vinci. This allows us to understand how drawing plays a fundamental role in the Leonardo da Vinci studies [Argan (1966)].

In this context we can understand that the true language of Leonardo da Vinci is drawing. Therefore, he dedicated specific efforts to make drawing a universal language for his studies. Of course, drawing is fundamental for an artist, moreover for Leonardo, it was fundamental in his studies regarding nature and the human body, as well as in his works as architect, civil engineer, and industrial engineer.

With the reference to the last item, he actually invented the technical design. For Leonardo da Vinci, the technical design was art. In fact, it comes from the Greek word $\tau\varepsilon\chi\nu\eta$ (transliterated as techne) which means *art*. The technical design skills in the projects on his machines were a new technical tool in order to represent mechanical equipments. The list of particular machines of Leonardo da Vinci are reported as follows:

- flying machines;
- time machines;
- lever cranes;
- mechanical musical instruments;
- civil engineer projects;
- transformation of motion;
- war machines.

Most of the Leonardo machines has been effectively realized not in the time of Leonardo but decades later, thanks to the clarity of his projects. Therefore, the Leonardo da Vinci mind was driven by a dynamical simulative organization able to express the results of his cognitive processes by using a new tool of technical design.

Even if Leonardo da Vinci used the perspective technique in his paintings, for example in the "Ultima Cena", or in the painting "L'Annunciazione", in the technical design he introduced axonometry. He usually introduced the third dimension in his machine designs, this was intended to give the perception of the dynamical behavior of the machine. He used the 2D projections in order to give more details of the complex apparatus.

The Leonardo technical design is a unique example for related work in his mind, his eye and his hand, giving an exact perception of the dynamical behavior of the machines like that we receive today by using advanced 3D

Fig. 4.2 Leonardo technical design skills on the project of the lever crane.

technical design tools. This is clearly identified by the drawing seen in Fig. 4.2, in which Leonardo provides what we now call an exploded-view of his design, a lever crane.

Quoting again Leonardo da Vinci: "Principles for the Development of a complete Mind: Study the science of art. Study the art of science. Develop your sense, especially learn how to see. Realize that everything connects to everything else."

Chapter 5

Leonardo da Vinci: civil engineer, architect and urban planner

5.1 Remarks and discussion

Our opinion is that many of Leonardo da Vinci mechanical machine projects were conceived as a result of his strong intention to realize his revolutionary ideas in civil engineering and architecture, consolidated to the modern concept of city planning [Heydenreich (1952)].

The city design of Leonardo da Vinci was revolutionary both in the organization and structure, marking the difference from Medieval City to modern Renaissance City. The Leonardo idea was to start from the general master plan and the drawing of the city. A classical example is the regulatory plan of Florence, based on a dodecagonal perimeter (Windsor 12681), on the diversion of the Arno river, and on a network of large meshes of primary orthogonal roads that can be viewed as a proposal for a rational model with respect to the disorder of the not-so organized Medieval City.

In order to perform the city planning, Leonardo da Vinci studied existing cities by using innovative methods like cartography. He used mathematical tools and geometric models. He used cartography to analyze the territory in order to relate the various elements that allowed him to design the new city. His studies consist of two types of designs: the sketches (typical of Leonardo) and the real topographic maps.

He realized the various maps by using also measurement and position instrumentations like the "bussola" (compass) and the "traguardo". An outstanding example is the map of Imola, shown in Fig. 5.1, that indicates the main key points of Leonardo's efforts in this area. Thanks to his design and drawing techniques, he gave precise details and a complete view, introducing the beauty of cartography.

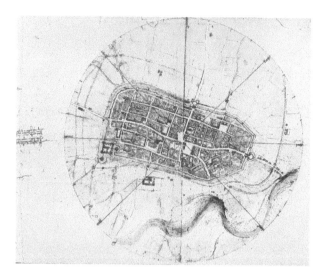

Fig. 5.1 Leonardo map of Imola, a small town near by Bologna, Italy.

In [Baratta (1912)] an exhaustive discussion on the innovative concepts of the Leonardo cartography is reported. A new way of interpreting cartography, a new method of deriving it, a scientific way to make maps characterize Leonardo da Vinci city maps. Moreover, the beauty of the maps that makes cartography like painting is seen.

These studies of Leonardo da Vinci was incorporated into the design of a rational city, and therefore there is a strict correlation of more disciplines in accordance with the Leonardo da Vinci thinking. The Leonardo da Vinci view of the modern city completely concurs with the viewpoint of his modern "thinking in complexity".

Leonardo da Vinci's single projects, as in buildings, places of worship, factories or markets and so on, had to be conceived in a particular context and, firstly, they had to be organized together in such a way as to assure the functional volumetric proportions within the correct area of the city. This concept was strictly oriented to assure the good life of the people in the city, the efficient communication and a network of modern services. Leonardo da Vinci must be considered the first modern urban planner in history [Firpo (1971)].

Indeed the principles of concreteness implemented in the Leonardo da Vinci projects are related, first of all, to the political and economic viewpoints. During the period 1484–1485, about one third of the people living in Milan, 50,000 persons, died due to the plague. He immediately did a comparison between Florence and Milan. In his mind, arose an aversion against the Medieval City: anarchist, disordered, filthy, made by temporary buildings, with people suffering from an unhealthy environment.

In order to free the people from the hard living situation of the city, he suggested a quite radical solution to make a new city. To this aim, the following actions were necessary: disembowelling, restoration, making spacious streets and courtyards, providing people with running water, building structures for hygiene, introducing decentralization of popular neighborhoods in rural areas, making a presentable city by building numerous stately homes. All these concepts appear in a draft letter that he wrote to Federico il Moro (Codex Atlanticus 65, v. b 1497). They are the palimpsest of the ideal city.

Fig. 5.2 Leonardo conception of ideal city structured over different specialized layers.

A draft of a piece of the ideal city of Leonardo is seen in Fig. 5.2. The main structural idea of Leonardo da Vinci is a multi-level city. The level of the water, where canals are allocated, that might be used for transportation of goods, to favor hydraulic exchange for cleaning the city and for addressing rain. The level of streets, that should assure people communications. The level of the buildings, where he conceived the rational assembly of more elements, such as those in the drawings in Fig. 5.3.

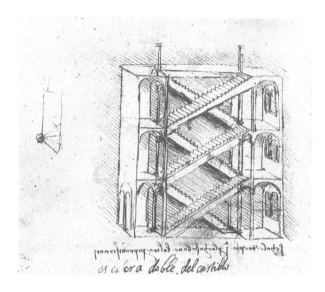

Fig. 5.3 Leonardo designs to connect different specialized layers by means of stairs.

Therefore, the main engineering strategies in order to realize a modern city require:

- machines for disassembling old cities;
- machines that help people to build;
- machines to be used for canal hydraulic flow control.

Indeed, the city projects by Leonardo da Vinci also involve the communication among the cities and efficient transportation. With such wide perspective, as he suggested in Florence with the Arno deviation in order to improve both the communication among cities and to use the water of the river for irrigation, Leonardo da Vinci started in 1482 the study to connect Milan with the Como lake and Milan with the Lago Maggiore. A glimpse of the level of detail adopted in the study can be seen in Fig. 5.4.

The study of river communication has been one of the most important topics for Leonardo da Vinci as a civil engineer. Here, the Leonardo da Vinci skills on hydraulics, turbulence, his perspective study on the modern city led to the conception of important hydraulic machines and mechanical apparatus addressing the control of river flows. In fact, the projects of the various "chiuse" (river locks) in order to obviate the unevenness of the land and to make navigation possible, required the conception of new mechanical

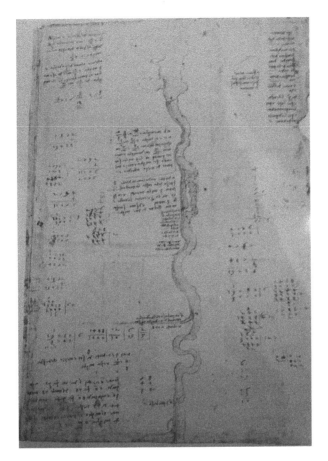

Fig. 5.4 Leonardo study of the Navigli river systems to connect Milan to the main lakes of northern Italy.

machines from Leonardo da Vinci, who invented them by coupling more simple machines, such as in Fig. 5.5.

Indeed the approach in managing the lakes and the river flows allows to conceive networks of adjustable level canals, a visionary idea for the economy at that time. The use of the hydraulic energy, the use of water as a resource for crops, the use of canals for transportation are outstanding projects from Leonardo. In order to perform these impressive activities, Leonardo again designed suitable mechanical machines.

The Arno deviation near Pistoia was also an impressive project of Leonardo da Vinci. He also had in his mind to build an artificial canal from Florence to Pisa to be dedicated to transportation that could also be

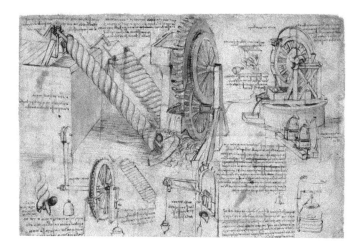

Fig. 5.5 Leonardo hydraulic studies for river locks.

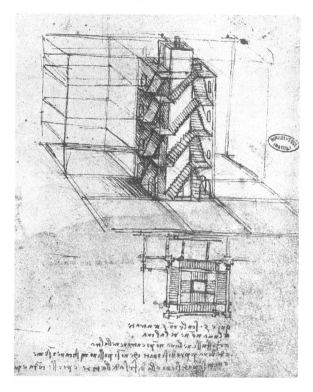

Fig. 5.6 Leonardo design on a multiple-stairs system.

very useful to prevent floods. In order to usefully control the flow of the various canals he conceived a set of "conche" (basins) and locks with the set of mechanical machines needed to control them.

Leonardo da Vinci was fascinated by the canals and when he went to Milan looking at the "Navigli" system, he dedicated more of his drawings to the particular structures that he wanted to realize to improve the system.

With the hydraulic machines that were conceived on the basis of previous considerations, Leonardo da Vinci machine designs arise from the need to build high-volume places and structures and to provide construction workers safe machines to optimize their efforts and alleviate their fatigue, as the stairs system in Fig. 5.6. The single structure design and the realization along with the organization in order to develop a continuous and

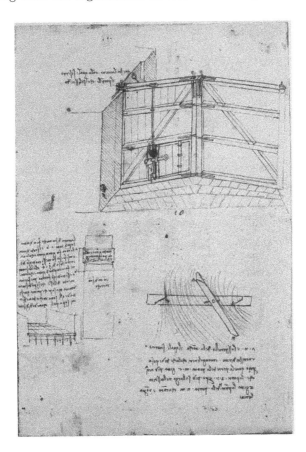

Fig. 5.7 Leonardo machines to move and lift objects.

fast process of city construction provided the input to design mechanical machines like those as in the drawing in Fig. 5.7.

Leonardo da Vinci designed also bridges, the self-supporting one is the most famous of his bridge, possibly conceived for military uses. Moreover, Leonardo conceived a bridge over Bosphorus in 1501, as reported in Fig. 5.8. The project has been realized in Norway in 2011 and will be realized in Istanbul on the Golden Horn site. It will be 200 meters long and 10 meters large.

Fig. 5.8   Leonardo project for a self-supporting bridge over Bosphorus.

Chapter 6

Leonardo da Vinci and the Engineering Scenario at His Time

6.1 Some notes

Leonardo da Vinci conceived the machines in a modern way: equipment for transforming and transferring energy. But which energy?

The electric power machines did not exist. Only human power, animal power, hydraulic and wind energy do appear useful in real life. The idea of saving energy for reuse was, therefore, particulary appealing. Moreover the unique autonomous motor that Leonardo could use in the design of his autonomous machines and robots was the spring-based motor. Leonardo da Vinci was clear in mind about the concept of saving energy into the spring and so, he knew how to use the elastic principle to save energy.

A lot of Leonardo's efforts were dedicated to develop equipment (machines) in order to use and to transform this energy in a simple manner and with high level of efficiency. During the time of Leonardo da Vinci, the engine based on heat energy had not yet been conceived, even if he realized the famous turnspit on folio 21r of the Codex Atlanticus, moved by air heated by a fire, and concentrated through the narrow neck of a chimney.

He had perspective insights on hydraulic energy and the possibility to transport it. In fact more of the Leonardo machines were conceived for this task. His attention was also strongly focused on wind energy. Besides, the real unique useful energy that Leonardo da Vinci considered in order to realize his bigger projects remained with the human power and animal energies. This is why the studies of Leonardo da Vinci in order to understand the living nature, his investigations for understanding the human muscle behavior, were also devoted to conceiving new motors and new innovative machines.

Indeed during the Leonardo da Vinci time, many materials were unknown. For example, the natural material from bamboo cane was unknown in Europe, while in China and in other parts of the world, it was considered as important as steel is nowadays, since it has the same mechanical properties but much more light. Indeed, new materials have only nowadays been used to realize some of the Leonardo flying machines, like the so-called *pinna di Leonardo* (Leonardo's wing), shown in Fig. 6.1 [Taddei (2007)].

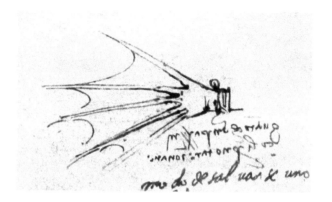

Fig. 6.1 Leonardo project for a bio-inspired wing.

The Leonardo da Vinci machines show that he had also achieved in his mind an ordered view of a new discipline: applied mechanics. He drew several prototypes of chains and bearings, as shown in Fig. 6.2, establishing a rigorous approach to applied mechanics. He clearly understood the importance of friction in designing equipment and he was able to conceive the first mechanical bearing.

Moreover, his designs show his clear knowledge on the kinematics of the rigid body and its application in achieving suitable mechanisms. The complete understanding and the scheduled order achieved in the Leonardo da Vinci mechanics are shown in his designs, including mechanisms with shapes and cans. Toothed wheels and gears were the key points of the Leonardo da Vinci mechanics, in particular in the application of flexible organs for lifting machines.

The applied mechanics of Leonardo da Vinci started from conceiving the simple machine principle. A simple machine, like the inclined plane, the lever, the wedge, the wheel and the axle, the pulley and the screw, is a mechanical device that changes the direction or the magnitude of the force. Moreover, thanks to the elementary machines principle, Leonardo da

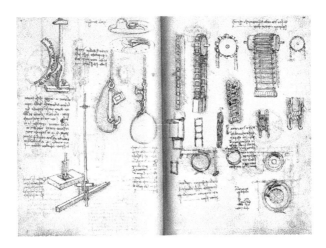

Fig. 6.2 Leonardo drawings about mechanical devices: chains and bearings.

Vinci conceived in an ordered way the designs of complex machines, also known as composite machines. Example of Leonardo composite machines are given in the following:

- unequal motion;
- rod transmission;
- alternative motion using a handle;
- spring with helical transmission;
- polishing motion;
- alternating rotator motion;
- reciprocating motion with blades and spit lever;
- pre-programmed motion along a track;
- gravitational gyroscope;
- self-blocking spring;
- progressive spring loading;
- out of phase spring powered motor;
- segmented reciprocating motion;
- belt transmission;
- movement along a pre-programming course;
- reciprocating direct motion with belt transmission;
- spring powered helical motor;
- examination of the connecting rod;
- flywheels with handles;

- multiple pulleys;
- wheels with no teeth.

Moreover, in the Leonardo da Vinci machines conception, he assumed the perspective that machines must also be considered as the element of a complex system, that is a plant made by driven machines and operating machines.

The inventions of Leonardo da Vinci emphasized the limitation of the mechanical equipment of that period and were strongly conditioned by the technological limits. Indeed the precise description of Leonardo machines allows us to identify their limits today: the artificial autonomous prime motor and its control. Analyzing the machine designs of Leonardo da Vinci, it can be conjectured that their behavior could be possible thanks to the fact that the real implementation was related to the imperfect equipment. In fact, in conceiving gears he took into account that mechanical tolerances are fundamental for the correct behavior of the equipment. This is only a simple example. But today imperfect systems [Fortuna *et al.* (2017); Bucolo *et al.* (2019)] and imperfections play a fundamental role in several topics of engineering.

Chapter 7
A Selected Number of Machines: innovation and control

7.1 The Leonardo da Vinci machines

The Leonardo machines are widely studied, proposed and realized all over the world. The various solutions adopted for the implementation of the equipment tell us that Leonardo's designs follow different approaches for their realization [Moon (2007)].

In the history of Leonardo studies, his machines were mainly relevant to the following classes of complex apparatus: war machines, like the robot soldier, the animal-mechanic machines, similar to the bird and the lion, the self-propelling cars, the automata and robots [Vinci (2019)].

Even if only written descriptions of the actual Leonardo machine automation do exist, specific designs of the automation apparatus did not survive. Indeed, the focus on automatic control problem in the Leonardo da Vinci designs is perceived by his ideas and, for different solutions feedback is solicited. Moreover, the lack of systems for handling energy, and for control, has limited the insight of Leonardo da Vinci studies on automatic control. If the apparatus and the machines for hydraulic applications can be revised, Leonardo da Vinci can then be considered as one of the fathers of Automation Engineering.

In the following, we will propose a set of composite machines focusing on the possibility of making them autonomous, both regarding energy and performed actions. Therefore, we have selected a number of Leonardo machines to include in them the modern engineering equipment and to realize completely autonomous machines. The selected Leonardo machines will be described and then the automatic control equipment for a specific innovation will be discussed.

Moreover, it is emphasized that the ideas of Leonardo da Vinci are stimulating and impressive, and give inputs to innovative solutions and outstanding demos.

In the next sections, the original models conceived by Leonardo da Vinci and realized following the guidelines of his projects by using mainly wood and light plastic material are presented. For each machine, the scheme of the additional mechanisms is described, providing suggestions on how to replicate our setups. Furthermore, the electronic devices included to control the machine behavior are described. Technical details on programming the microcontrollers and the related electrical schemes are given in Appendix B.

7.2 Drop propagation

We start our gallery from a comprehensive study on a machine able to simulate the wave propagation in a fluid when a drop hits the surface. This machine is not actually based on a Leonardo da Vinci drawing but several studies on hydraulics and wave propagation characterize the work of Leonardo. Leonardo da Vinci left a great heritage in the form of notebooks and loose sheets full of notes and drawings on a great variety of topics. Although an important part of them is unfortunately lost, it is undeniable that in the history of science, Leonardo was the first true researcher. Among his many works as artist, engineer and scientist, those concerned with fluid properties, flow and transport phenomena are surely the most original [Macagno (1985)]. In [Macagno (1991)] the author studied a few of Leonardo da Vinci experiments, that cover a wide range of situations in hydrostatics, they include the experiment of the floating double box [Macagno (1982)], which is crucial to old physics and Leonardo had difficulties in its implementation. Also in hydrostatics, the designs of experimental determinations of forces and pressure distribution on the walls of a tank are remarkable. In the area of flow phenomena, some of Leonardo's studies (impingement of jets, perforation of nappes by jets, siphon flow, and flow of granular materials) have been investigated in [Macagno (1991)]. The criterion for the above selection is that each experiment illustrates an important facet of the experimentalist's work, including both successes and failures. In the Leicester Code, Leonardo da Vinci formulated three different indexes of that treatise on water that will never be completed. It was proposed to reveal its basic structure, to illustrate the reasons for the cataclysms it produces, to investigate its whirling motions, to devise effective solutions to avoid the ruin of river banks, and to emphasize the similarities between water and air, between fish motion and birds.

Leonardo da Vinci strived to focus on the fundamental properties of the water element. He studied the behavior of the drop that constitutes its ultimate elementary component, highlighting its spherical shape, its cohesion with the rest of the water element (like iron to magnet) and the elasticity that is evident, elongating before falling, and then use the same method to analyze the dewdrop and the water bubble, stating that they support themselves because they have a perfectly hemispherical dome shape. Leonardo da Vinci, through the study of motion and water measurement, offered to write one of the first documents on the study of the propagation of oscillations due to the impact of a body on a body of water, as in the drawings in Fig. 7.1.

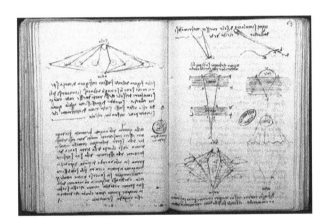

Fig. 7.1 Leonardo's Codex Leicester.

Starting from Codex Leicester many artists have approached some or all aspects of motion and water propagation, their work relates not only to the ideas of closer observation, but also relate back to Leonardo in some way. For example, some works include a series of strobe-light photos by MIT professor Harold Edgerton, one in particular called "Milk Drop Coronet," created in 1979, that depicts a milk drop landing on a shallow plate of milk; the splash creating a crown effect. These photos can be linked to Leonardo's drawings on the elasticity of water.

In particular, Harold Edgerton, through the use of photography techniques called milk drop coronets offered to freeze moment by moment the propagation motion of the surface waves due to the impact of a body on water. In this chapter, this phenomenon on the topic and drawn in turn from the studies carried out by Leonardo da Vinci has been investigated. Subsequently for educational purposes, an automata, with simple manual movement that represents a kinematic motion was realized. That is a machine which simulates in a synthetic but a spectacular way, one of the simplest manifestations of nature: a drop of water that falls on a stretch of water. The ripples of the surface of the water that are produced, for example, when a drop of rain falls into a puddle are two-dimensional waves: the shape of their wave fronts are called circular waves. No one before Leonardo da Vinci had tried to describe it in such detail comparable to the infinite motions of water, its interactions with other elements and the continuous modifications it produces on the surface and in the belly of the Earth. He had the conviction that water science would allow him to reveal the organization and functioning of the entire natural environment. But after first describing the automata, it is important to introduce the mathematical model of this phenomena.

Periodic phenomena follow our life. On the largest scale, mechanical periodicity is inherent in the movements of celestial bodies and of the Earth, producing secondary periodicities, which we recognize as the yearly seasons, alternating day and night, the ebb and flow of tides, and in many other ways. The importance of oscillatory phenomena facilitates the drop propagation model.

This model is dominated by the physical laws that describe its correct propagation along the three dimensions and approximate the frictions that separate the theoretical world from the practical one. In first approximation, we will look for laws of elementary harmonics to understand theoretically what happens to our system as time changes. In the second approximation, the physical system should be studied with the addition of damping oscillation, perhaps, otherwise periodic, due to forces of friction often unwanted yet to which we are subject perennially. Before dealing with the state equation of our system we try to understand its physical and theoretical aspects. We start with more elementary oscillation and then move on to the version that takes into account the forces of friction and the wave propagation as a function of time and space.

Begin the analysis with Newton's second law of motion. There is only one force — the restoring force of the spring (which is negative since it acts opposite to the displacement of the mass from equilibrium). Replace net force with Hooke's law. Replace acceleration with the second derivative of displacement.

$$ma(t) = -ky(t)$$
$$-m\omega^2 x(t) = -ky(t) \tag{7.1}$$
$$\omega^2 = \frac{k}{m}$$

Thus we note that the pulsation and therefore the period of the harmonic motion are determined by the value of the mass of the material point and the characteristics of the elastic pull force, due to dynamic conditions, on which the amplitude and initial phase do not depend. A material point experiences a simple harmonic motion if the hourly law is defined by:

$$y(t) = A\sin(\omega t + \phi) \tag{7.2}$$

where A, ω, and ϕ are constant and they are respectively amplitude, pulse and initial phase. The harmonic motion is a variable motion, with the position, velocity and acceleration being variable in time. The velocity and acceleration can be described by the following equations:

$$v(t) = \frac{dy(t)}{dt} = \omega A \cos(\omega t + \phi)$$
$$a(t) = \frac{dv(t)}{dt} = -\omega^2 A \sin(\omega t + \phi) = -\omega^2 y(t) \tag{7.3}$$

Through (7.3), we can obtain the characteristic equation for the harmonic oscillator:

$$\frac{d^2 y(t)}{dt^2} + \omega^2 y(t) = 0 \tag{7.4}$$

This is a second order, linear differential equation. We have two possible functions that satisfy this requirement — sine and cosine — two functions that are essentially the same since each is just a phase shifted version of the other. The generic solution will be:

$$y(t) = a\sin(\omega t) + b\cos(\omega t) \tag{7.5}$$

By placing $A = \sqrt{a^2 + b^2}$ and $\tan\phi = \frac{b}{a}$, it follows that $y(t) = A\sin(\omega t + \phi)$. The dynamic state of this system is related to y that varies with harmonic law. If the harmonic oscillator is blocked by an inertial force the law of motion is:

$$ma = -ky - \lambda v$$

$$m\frac{d^2y(t)}{dt^2} = -ky(t) - \lambda\frac{dy(t)}{dt} \tag{7.6}$$

$$\frac{d^2y(t)}{dt^2} + \frac{\lambda}{m}\frac{dy(t)}{dt} + \frac{k}{m}y(t) = 0$$

where $\gamma = \frac{\lambda}{m}$ is the damping coefficient and $\omega_0 = \sqrt{\frac{k}{m}}$ is the natural frequency. Therefore, the equation of motion can be write as:

$$\frac{d^2y(t)}{dt^2} + 2\gamma\frac{dy(t)}{dt} + \omega_0^2 y(t) = 0 \tag{7.7}$$

This equation is called the differential equation of the damped harmonic oscillator; it is the most complete example of the linear differential equation of the second order, with constant, homogeneous coefficients. Note that if it exists, the solution $y(t)$ is proportional to $e^{\alpha t}$.

$$\frac{d^2 e^{\alpha t}}{dt^2} + 2\gamma\frac{de^{\alpha t}}{dt} + \omega_0^2 y(t) = 0$$

$$\alpha^2 e^{\alpha t} + 2\gamma\alpha e^{\alpha t} + \omega_0^2 e^{\alpha t} = 0 \tag{7.8}$$

$$e^{\alpha t}(\alpha^2 + 2\gamma\alpha + \omega_0^2) = 0$$

$e^{\alpha t}$ is a solution only if α satisfies the condition $\alpha^2 + 2\gamma\alpha + \omega_0^2 = 0$, therefore $\alpha = -\gamma \pm \sqrt{\gamma^2 - \omega_0^2}$.

There are three possible conditions:

- Hard damping, i.e. $\gamma^2 > \omega_0^2$
- Critical damping, i.e. $\gamma^2 = \omega_0^2$
- Weak damping, i.e. $\gamma^2 < \omega_0^2$

The solution depends on the relation between the oscillator's physical parameters. Hard and critical damping do not permit oscillation so only the condition of weak damping has been investigated. In this case, the damping coefficient is less than the natural frequency, leading to $\alpha^2 < 4mk$. The solutions are complex and conjugate so that one has $y(t) = A_0 e^{(-\gamma t)}\sin(\omega t + \theta)$. The initial conditions determine the initial amplitude and phase.

In weak damping condition, the material point will move with an oscillation pulse $\omega = \sqrt{\omega_0^2 - \gamma^2} < \omega_0$. The damping coefficient $e^{(-\gamma t)}$ is very important to describe the phenomena of drop propagation. The object falls into a body of water carrying the water level at the point of impact to a width equal to $-A_0$, and at that point then the water oscillates as described above, following the equation of harmonic motion slightly damped, reducing the amplitude of the oscillations according to the time elapsed.

In reality there is always friction and this results in a free oscillation always damped. It is important to study how to make the oscillation persistent, therefore how to realize a real physical system that oscillates with defined frequency and constant amplitude, even in the presence of these forces of viscous friction. Applying to the equation of motion a sinusoidal force, we then get:

$$ma(t) = -ky(t) - \lambda v(t) + F_0 \sin(\omega t)$$
$$\frac{d^2 y(t)}{dt^2} + 2\gamma \frac{dy(t)}{dt} + \omega_0^2 y(t) = \frac{F_0}{m} \sin(\omega t) \tag{7.9}$$

The equation obtained is no longer homogeneous. Also the force F has got a pulse ω which is generally different from ω_0 of the oscillator. We are looking for a special non-damped oscillatory solution of type $y(t) = A\sin(\omega t + \phi)$, that is with the same pulsation of the impressed force. If such a solution were allowed and exist, then the more general solution would be $y(t) = A\sin(\omega t + \phi) + ae^{\alpha_1 t} + be^{\alpha_2 t}$. We note that damping the transient phenomenon in a time that is characterized by the damping coefficient γ, would remain as permanent oscillation due to the force impressed. Thus inserting the solution at speed to the equation (7.9) we obtain:

$$[(\omega_0^2 - \omega^2)A\cos(\phi) - 2\gamma\omega A\sin(\phi)]\sin(\omega t) + [(\omega_0^2 - \omega^2)A\sin(\phi)$$
$$+ 2\gamma\omega A\cos(\phi)]\cos(\omega t) = \frac{F_0}{m}\sin(\omega t) \tag{7.10}$$

It follows

$$A = \frac{F_0}{m} \frac{1}{\sqrt{(\omega_0^2 + \omega^2)^2 + 4\gamma^2 \omega^2}}$$

and

$$\tan(\phi) = -\frac{2\gamma\omega}{\omega_0^2 + \omega^2}.$$

The harmonic oscillator responds to a sinusoidal stress with a sinusoidal displacement, out of proportion to the force of an angle ϕ, whose pulsation is not the proper ω_0, but is equal to the desired ω, imprinted by the external force. Width and depth depend on the value of the pulsation ω and not on the initial conditions that affect only the constants a and b of the transient part.

According to the value of the pulsation ω we have three possible conditions:

- $\omega \ll \omega_0$, so that $A \approx \frac{F_0}{k}$, $\phi \approx 0$, and $y(t) = \frac{F_0}{k} \sin(\omega t)$;
- $\omega = \omega_0$, so that $A \approx \frac{F_0}{2m\gamma\omega_0}$, $\phi \approx \frac{\pi}{2}$, and $y(t) = \frac{F_0}{2m\gamma\omega_0} \cos(\omega t)$ (resonance);
- $\omega \gg \omega_0$, so that $A \approx \frac{F_0}{m\omega_0^2}$, $\phi \approx \pi$, and $y(t) = \frac{F_0}{m\omega_0^2} \sin(\omega t)$.

For simplicity of construction our model presents a phase response with the applied force. Take note of the equation of motion that the model has investigated, it is important to study the wave propagation. In fact, in our system, we have a series of crests and wave bells that propagate circularly creating different concentric wave fronts that are damped based on time and propagation distance. Such waves are called surface waves, they are propagated by oscillating vertically along the interface between two fluids with different densities, but they do not propagate either as longitudinal waves or as transverse waves.

To calculate the surface wave velocity, we consider two cases. Low water, that is when the depth of the water h is much less than the wavelength $\nu = \sqrt{gh}$, and deep waters, that is when the water depth h is greater than the wavelength $\nu = \sqrt{\frac{g\lambda}{2\pi}}$.

The speed of the waves in shallow waters therefore depends only on the depth h and the gravitational acceleration g. While in deep waters the speed of the wave depends on the gravitational acceleration g and the wavelength λ. The perturbation of a field (by field we indicate a physical magnitude that can be defined at any moment in each point of space) produced by a source, that propagates in space is represented by the function $\xi(x, y, z, t)$ called the wave function. Our particular situation is constituted by the so-called plane waves, described by function $\xi(x,t)$, spatially one-dimensional, that depends on the space coordinate x alone as well as time. Where in our system x would be the radius that locates the set of points in the same state. The plane waves equation is a differential equation, in particular:

$$\frac{\partial^2 \xi(x,t)}{\partial x^2} = \frac{1}{\nu^2} \frac{\partial^2 \xi(x,t)}{\partial t^2} \qquad (7.11)$$

which is called D'Alambert equation and where ν^2 is the propagation velocity. This equation is a differential equation to the partial derivative of the second order, homogeneous, at constant coefficients, linear in the unknown function ξ. The solution functions of this equation can be of any kind but the dependence on the variables x and t must assume one of the two forms $\xi(x-\nu t)$ or $\xi(x+\nu t)$. The argument of ξ must therefore contain the variables x and t in the form of a linear combination. The most general solution is thus obtained as a linear combination of the previous two solutions as $\xi(x,t) = \xi_1(x-\nu t) + \xi_2(x+\nu t)$. The physical significance of the functions $\xi_1(x-\nu t)$ and $\xi_2(x+\nu t)$ lies in the fact that they represent, because of their structure, a phenomenon of propagation along the axis x with velocity ν. The value assumed by ξ_0 at time t_0 and in the position x_0, $\xi_0 = \xi(x_0 - \nu t_0)$, finds itself anytime $t > t_0$ at the point x that satisfies the condition $x - \nu t = x_0 - \nu t_0$ as $x = x_0 + \nu(t-t_0)$.

Reasoning, valid if we do not consider the effect of damping due to the viscous friction of the liquid, and which expresses a uniform rectilinear motion along the x-axis with velocity v. Continuing to neglect the effect of viscous friction, we now seek a solution to the D'Alambert equation for harmonic waves.

The wave function for harmonic waves is $\xi(x,t) = \xi_0 \sin k(x-\nu t)$, where ξ_0 is the amplitude and k is the wave number. It can be also expressed as $\xi(x,t) = \xi_0 \sin(kx - \omega t)$, where $\omega = k\nu$. It is called harmonic wave pulse. If we set a certain instant t_0, the value of the wave function repeats itself as a function of the variable x, so there will be two points so that $k(x_2 - x_1) = 2\pi$ for each equidistant pair of consecutive points. The distance $\lambda = x_2 - x_1$ is called wavelength and is calculated as $\lambda = \frac{2\pi}{k}$.

If instead we set x in a given position x_0 the value of the wave function repeats itself as a function of the variable t, therefore there will be two instants such that $\omega(t_2 - t_1) = 2\pi$. The time interval $T = t_2 - t_1$, is the wave period and is calculated as $T = \frac{2\pi}{\omega}$. Putting the two results together we get that $\lambda = \nu T$ and equivalently $\nu = \lambda f$, where f is the damping frequency. We can therefore conclude that the system representative of the phenomenon of wave propagation on a body of water can be mathematically schematized with the equations of propagation of plane waves in circular form.

Starting from Leonardo's study and Edgerton's opera, an automata has been realized, with simple manual movement that represents a kinematic motion. That is a machine which simulates the wave propagation. We can decompose this automata into four main components:

(1) Structure;
(2) Cam Shaft;
(3) Piston rods;
(4) Concentric rings.

The structure made with plywood panels to support the camshaft, which must rotate inside it and by means of two overlapping panels spaced within a few centimeters apart, aligns the connecting rods, whose head pushes the numerous rings present above this structure. The Cam shaft represents the heart of this automaton. Equipped with an axial symmetry, he who interacts directly with the gear motor transforms the rotation of its axis into the kinematics of the connecting rods. This consists of a series of carefully dimensioned and positioned shoulders that allow the creation of an extraordinary visual effect that simulates the propagation of circular waves whose wavelength, i.e. the distance between two bellies or between two ridges, is set for the construction of this component. Let us note from Fig. 7.2 how the shoulders of the central tree are larger than those present at the sides and how they gradually get smaller and smaller, simulating a viscous damping of a weak type. Another point to highlight with Fig. 7.2 is how the wavelength is fixed and visible to the naked eye, made evident by looking on either side the right or the left, with the presence of the two bellies fixed by construction at a constant distance.

Fig. 7.2 Cam shaft of the drop model.

The piston rods are the most difficult components to implement correctly (see Fig. 7.3). These connecting rods have the task of transforming the rotary motion of the camshaft into oscillatory motion of our interest.

Fig. 7.3 Piston rods of the drop model.

Despite their intuitive operation, they cannot be immediately implemented, due to their continuous synchronous movement and the numerous frictions that they are subjected to during operation. There are 19 in total and if one of them did not flow well, either because of the camshaft or because of the structure that held them in axis, it would be cause enough for the immediate stop of the global kinematics.

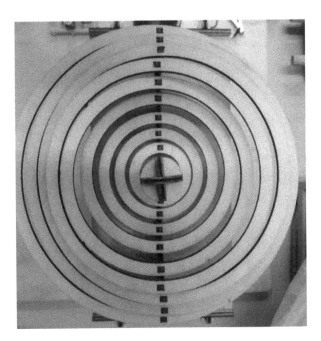

Fig. 7.4 Concentric rings of the drop model.

The concentric rings, shown in Fig. 7.4 are placed on pairs of connecting rods, two to two having the same distance from the centre, they represent the circular wave and the oscillation dictated by the joint kinematics of camshaft and connecting rods, simulating the propagation of the wave.

As we can see in Fig. 7.5 the camshaft was moved by a motor that in this case is a DC motor with a reduction mechanism in order to decrease the number of revolutions but increasing the torque necessary to move the system mentioned, as shown in the details in Fig. 7.6.

Fig. 7.5 Automated drop model.

Fig. 7.6 DC motor with increased torque adopted to actuate the drop model.

Table 7.1 Input Voltage V_{in} of the DC motor versus number of revolutions (RPM) per minute

V_{in} [V]	RPM
6	25.1
6.5	27.7
7	30.3
7.5	33
8	35.5
8.5	38

In this regard, a high number of revolutions per minute would have been detrimental to our purpose because the system for construction was not designed to exceed a threshold of around 50 revolutions per minute. The choice of the gearbox was therefore forced by the demand of our system to possess a high torque with a low number of revolutions. In the table, we report the motor voltage input and the corresponding number of revolutions per minute at the exit of the engine with the load attached.

We note that in the range considered $(6 - 8.5V)$ there is an almost linear increase in the number of revolutions per minute as a function of the input voltage, such linearization allows us to implement an open loop control on the frequency of oscillation of our system that depends on the input voltage. The linearization developed in the range between 6.5V and 8V is as follows:

$$RPM = 5.2V_{in} + 6,1 \qquad (7.12)$$

The automation solution envisaged for this project consists of the realization of a complete system of control and management of the camshaft rotation of the model under examination, based on the Arduino UNO® microcontroller.

Through a transistor NPN $BC547$ the voltage to be supplied to the gear motor to vary its speed is managed. Note the presence of silicon diode $1N4007$ connected in parallel to the power terminals of the gear motor: the use of this component is essential since the deactivation of an inductive load, such as an engine, results in voltage transients, or spikes that can cause the pilot transistor to rupture. The reason for creating this transient voltage is that the voltage at the ends of the inductive load is proportional to the value of the inductor and the speed with which the current varies. During the transition of current from a constant value to a zero value, or from a state of active loading to deactivation, the current generally varies very quickly, causing an increase in tension at the ends of the inductor. This voltage is found right at the head of the pilot transistor and can easily exceed the breakdown voltage that leads to the breakage of the transistor. A 10kΩ potentiometer has been used to adjust the speed and control the ignition system, through which goes a portion of the voltage of 5V, measuring the latter through the microcontroller's $A0$ analog input, so as to manage the output voltage at the pin "driver" PWM5 (section gear motor drive). The use of a button normally open, with a pull-down resistance of 100 KW, allows to start and stop the operation of the model. To complete the user interface of the entire automation system is the LCD1602 — a 16×2 display which, connected with the microcontroller, makes it possible to visualize the status of the system, the value of the voltage sent to the gear motor and the corresponding value of RPM. The complete setup is reported in Fig. 7.7.

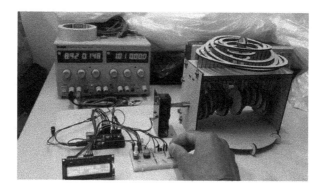

Fig. 7.7 The automated drop model.

Through the hardware realized and the developed software, it is possible to record the frequency of the wave according to the supply voltage of the gear motor. This frequency is related to the different objects that impact on the body of water. It is possible that a feedforward control can be related to the frequency of the wave with the DC motor velocity through Arduino. The results have been shown in Fig. 7.8.

7.3 Mechanical clock

In history, these type of machines have provided a lot of impressive examples. Moreover, even if they are intended as the equipment that allow an automatic "time detection", some people, before Leonardo, starting from

A Selected Number of Machines: innovation and control   57

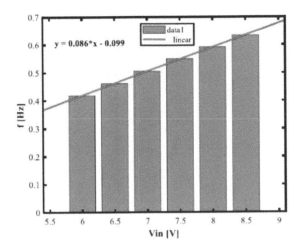

Fig. 7.8   Wave frequency versus Input Voltage provided to the DC motor.

Vitruvio, conceived the time machines as limiting to clockwork mechanisms; the Leonardo da Vinci clockwork is included in his wide spectrum of research projects.

A very appealing time machine of Leonardo is that which appears in the Manuscript B, Folio 33V, 34 R. It has never been deciphered, and realized only recently [Milano (2019)]. Let us imagine that the boxes are supplied by water, but it could be another fluid. Moreover, the water could continuously flow, this is in principle a perpetual motion machine that never stops working, this is a prototype of time machine.

The concept of time was strongly perceived by Leonardo da Vinci who tried to explain time by using design and machines conception. Therefore he worked to both face the concept of time from a philosophical viewpoint and also from a technological viewpoint. In the image in Fig. 7.9(a), the original model of the Leonardo project is displayed. He used an iron-based system to get the clock charge and a weight system in order to balance the clock.

The innovation project based on the model of the Leonardo clock consists of intelligent devices to charge the clock when the weight reaches the lowest position. To accomplish such a task, opportunely located sensors and an actuation system are necessary. Moreover, a microcontroller can be included to drive the charge process. We adopted a pair of reflective optical sensors with transistor output, namely the TCRT5000 produced by Vishay

(a)
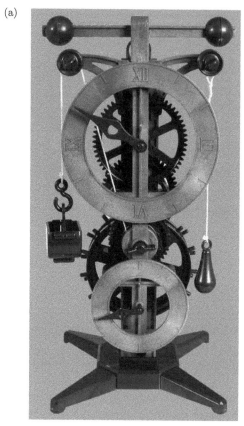

(b)
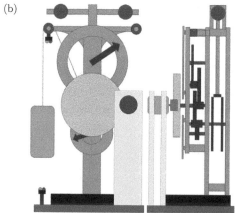

Fig. 7.9 Realization of the Leonardo clock: (a) original model, (b) scheme of the automated setup.

A Selected Number of Machines: innovation and control 59

(a)
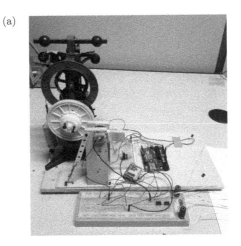

(b)

(c)
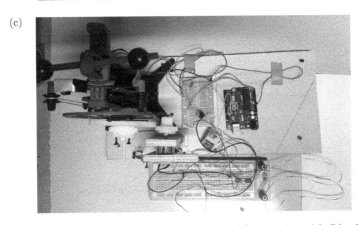

Fig. 7.10 Pictures of the automated Leonardo clock: (a) original model, (b) scheme of the automated setup.

Semiconductors. This device is very compact in size ($L \times W \times H$ in mm: $10.2 \times 5.8 \times 7$), robust in operation and, also contains the infrared photodiode emitter and the detector phototransistor. The latter is provided with a filter block for the ambient light in the visible range. To actuate the charge system we adopted a controlled 3V DC motor on whose shaft is spliced a gear of 16mm diameter. Such a gear contacts a larger gear, with 87mm diameter in order to multiplicate the torque, which is directly spliced on the circular spring allowing for clock charge. The actuation is governed by an Arduino® microcontroller, which is interfaced with a relay to deactivate the motor during the clock operations. The schematic of the innovative machine is seen in Fig. 7.9(b), while several pictures of the automated clock can be seen in Fig. 7.10.

(a)

(b)

Fig. 7.11 Realization of the Leonardo lever crane.

7.4 The Lever Crane

The Lever Crane was realized thanks to the Leonardo designs that had been conceived for a system able to lift weight. The Lever Crane system uses a ratchet mechanism that rotates in one direction to lift a weight with a fraction of the strength of a simple pulley. The original sketches are included in Fig. 4.2. It was realized by the model displayed in Fig. 7.11(a).

The automatic model of the Lever Crane is designed in order to provide a continuous motion of lifting and lowering the weight. Switching between the two directions is ensured by small push buttons located at the end of the run of the lever. Actuation is obtained equipping the model with a servo-motor spliced with a gear which contacts a larger gear located on the crane axis. The schematic representation of the automated lever crane can be seen in Fig. 7.11(b). Moreover, Arduino® controls the motion reacting to external input provided through a bluetooth module HC-05. The complete setup is shown in Fig. 7.12.

7.5 Rolling ball timer

The Rolling ball timer is based on a sketch of an escapement mechanism drawn by Leonardo da Vinci. The system in based on exploiting the principle of gravity as kinetics to measure the flowing of time. This is another exhaustive example of the attention of Leonardo da Vinci to the time concept: the rolling ball timer was conceived to fix different time intervals. The ideas of building clock-based systems, and therefore intelligent systems, were recurrent in Leonardo da Vinci's mind.

The realization of the Rolling ball timer is based on the model in Fig. 7.13. The working principle is based on a lever supporting a weight, which falls down each time the rolling ball completes its path. The ball then mechanically actuates a rod which inverts the balance of the board and gets the ball rolling again. We can introduce in this scheme a system devoted to bring the lever up to the starting position when it reaches the lowest point, allowing for a continuous flow of temporal intervals. The automation is realized through a DC motor shafted to a gear of 25mm which contacts a larger gear with a diameter of 70mm. This latter gear is coaxial with a bullwheel over which a rope is wrapped. The rope is linked to the timer lever. When the lever reaches the lowest position, it hits a switch which drives the Arduino® board. An electromagnetic actuator ensures

(a)

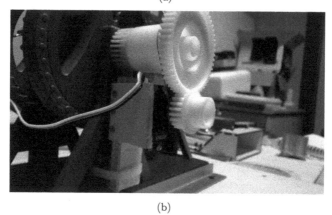

(b)

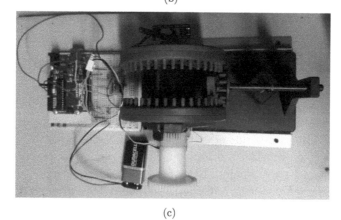

(c)

Fig. 7.12 Pictures of the automated Leonardo lever crane.

A Selected Number of Machines: innovation and control 63

Fig. 7.13 The rolling ball timer.

that the rods are not blocking the board while the lever reaches the upper position.

The complete innovated model is as in Fig. 7.14, where details of the actuation systems are provided.

7.6 Mechanical drum

The mechanical drum is reported in a sketch of Codex Atlanticus, Folio 837. It was conceived as an entertainment machine by Leonardo da Vinci in the courts. In fact, it is included in the set of Leonardo machines categorized with music.

The machine is a cart equipped with a mechanical drum. When pulled or started by a handle, the gears rotate the two sides of the drum which is equipped with pegs. These pegs, which can be placed in various positions, move their sticks that beat the large drum at the back. Changing position of the pegs alters the rhythm of the music. The innovative mechanical drum includes some driving equipment electrically controlled with autonomous power supply and a remote controller. The model with the control setup is seen in Fig. 7.15(a).

It consists of a wheeled cart supporting the control, the actuation and the power supply devices, linked to the mechanical drum. For the actuation system, we adopted a DC motor with multiple reduction gears in order to provide the wheels with the torque necessary to overcome friction and start

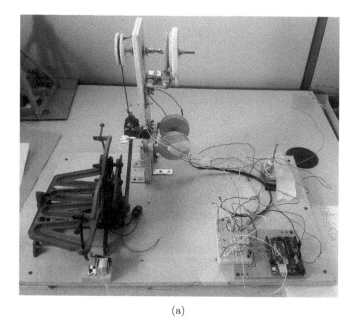

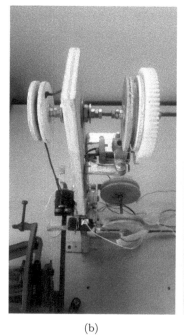
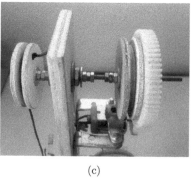

Fig. 7.14 Pictures of the automated Leonardo rolling ball timer.

A Selected Number of Machines: innovation and control 65

(a)

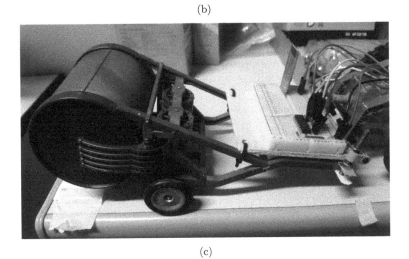

(b)

(c)

Fig. 7.15 Pictures of the automated Leonardo mechanical drum.

moving and playing the drum. A zoom on the gear system is seen in Fig. 7.15(b). Moreover, a servomotor is used to deflect two coordinated wheels in order to allow turnings. The model is remote controlled by using a bluetooth device linked to a smartphone.

7.7 Flying Machine

Leonardo was obsessed with building a Flying Machine. He conceived more machines, some of them bio-inspired and others that consider human power and applied mechanics as the main ingredients of the design. Indeed, as a pacifist, Leonardo da Vinci had an internal dilemma: between the scientific interest in the Flying Machine and its possible application for military usage. This has been a debate in the 2002 BBC Television series on Leonardo and it is till now a lesson for modern times from Leonardo da Vinci, a great engineer but also a visionary man from social and historical aspects too.

A model realized in wood from the design of Leonardo flying machine was used. We included in the automated version of the flying machine a high torque DC motor which actuates a rotating gear that is linked to a rope at an eccentric position. Motion is transferred by means of two bullwheels to the base of the flying machine which moves upward and downward alternatively. The automated setup is seen in Fig. 7.16, where specifies of the actuation system are shown (Figs. 7.16(b) and 7.16(c)).

The effect of motion is here reproduced by comparing the still flying machine in Fig. 7.17(a) to the picture taken by using a long exposure camera as in Fig. 7.17(b).

7.8 Self-supporting bridge

The design of this project is included in the Codex Atlanticus, Folii 69 AR, 71 V. The sketch of the Leonardo design is seen in Fig. 7.18.

Leonardo designed this bridge under the patronage of Cesare Borgia. It requires no nails or ropes to hold it together, as the bridge is self-supporting. The Leonardo bridge was realized as in Fig. 7.19. Moreover, it was used as an experimental platform for dynamic vibrational studies. An electromagnet was located on the bridge in order to provide a vertical oscillation of the structure. The frequency of the oscillation was varied using a signal generator in order to derive an experimental evaluation of the frequency response of the self-supporting bridge. Interestingly, the model of the Leonardo bridge proved to be highly robust to a wide range of low and

A Selected Number of Machines: innovation and control 67

(a)
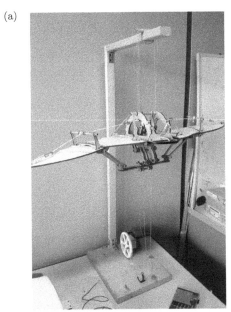

(b)

(c)

Fig. 7.16 Pictures of the automated Leonardo flying machine.

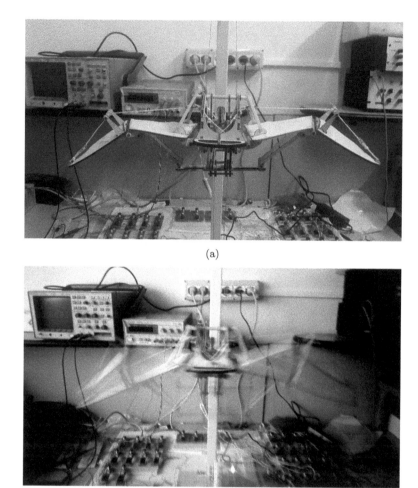

Fig. 7.17 Motion of the automated Leonardo flying machine.

high frequencies but tends to disassembly around the resonance frequency of the structure, found in our experiments to be around 10Hz.

7.9 Transformation of motion

The problem of transferring motion over different planes of actuation is crucial in the Leonardo da Vinci studies. Given the limited number and efficiency of the available sources of energy during his time, Leonardo pro-

A Selected Number of Machines: innovation and control 69

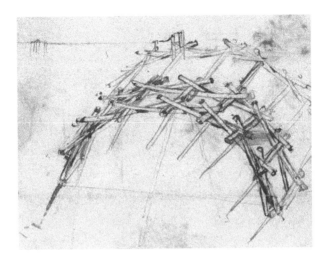

Fig. 7.18 Study for a self-supporting bridge.

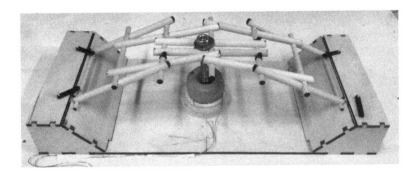

Fig. 7.19 A self-supporting bridge for dynamic vibrational analysis of the structure.

posed several strategies to transform and transfer a rotation parallel to the ground, such that indicated by an ox carrying a lever connected to a wheel, with a rotation orthogonal to the ground, required to lift or lower a weight. His studies led to different solutions, either based on gears, as in Fig. 7.20(a) or on bullwheels and ropes, as in the drawings in Fig. 7.20(b).

We realized a completely autonomous system of bullwheels able to transfer energy from a standard DC motor to an additional motor acting as a dynamo lighting a simple LED. In the schematic representation seen in Fig. 7.21, the motors are located in separate positions and the transformation of motion leads to actuate the dynamo. The realized model as in

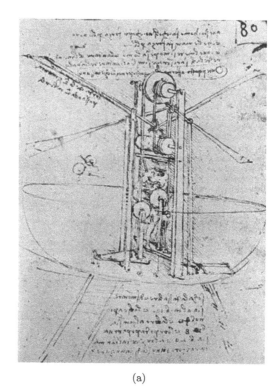

(a)

(b)

Fig. 7.20 Leonardo studies for transforming motion.

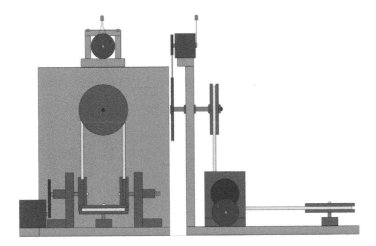

Fig. 7.21   Schematic of the setup for the transformation of motion.

Fig. 7.22 indicate several views and specifies of the machine. In particular, the shaft of the actuated motor is spliced to a gear contacting a larger gear joint with a bullwheel. An elastic rope allows for the rotation of the other bullwheels, especially a joint with a further larger one. This bullwheel motion is then transferred to the DC motor acting as a dynamo.

The whole system can be controlled through a smartphone linked to the bluetooth device installed on the microcontroller.

7.10   Final considerations

The study on Leonardo da Vinci has been inspired by a set of activities in which we have been involved in order to celebrate the 500th anniversary of his death. Even if we collected the material, the work has been possible thanks to the passion of more people. We want to remark that many of the presented machine's models and their electronic equipment have been realized by the students. Most of the material used in the projects was derived from disposed equipments.

From our studies on Leonardo da Vinci, we learned a lot of things. The first is that, even now, Leonardo da Vinci's works inspire and push new ideas and new projects, even in a modern and global era. An overview of the Leonardo da Vinci studies, shows that he really conceived the System Science approach, probably for the very first time in such a systematic way. In our opinion, the era of Complexity was born with Leonardo da Vinci.

Fig. 7.22 The transformation of motion: (a) complete setup, (b) especially for the motor actuation, (c) especially for the dynamo.

Chapter 8
Epilogue

The preparation of the items contained in this book gave us several research ideas, proving how strong is the cultural message of Leonardo da Vinci, even today. In particular, the flying machine and the experiments described in Chapter 1, pushed us to propose a novel project involving a wide range of scientific fields.

Our project, called *Labonstrato*, at the Dipartimento of Ingegneria Elettrica Elettronica e Informatica of the University of Catania, aims to organize a small laboratory for experiments in the stratosphere, up to 100000 feet, based on the use of helium weather balloons. The project involves different research centers in Italy interested in different topics but joined by the need for low-cost experiments in stratosphere conditions. In Fig. 8.1(a) we report the box conceived to securely host different experiments while flying in extreme temperature and pressure conditions, exposed to cosmic rays, UVs, ionizing radiations and high concentrations of ozone. In 2020, the first complete flight with experiments will be performed.

Thus, after five hundred years since his passing, Leonardo da Vinci continues to stimulate new projects. And his flying machine may reach new altitudes.

This book is a small contribution to his stimulating ideas. First of all, starting from the Leonardo da Vinci anniversary, a group of students from the Master Course in Automation Engineering and Control of Complex Systems at the University of Catania started to study the control problems from a new point of view. The science is built in time, to go on, to stimulate new generations, joining the past with the present time and to create a project for the future age without the need to know the history. And always bearing in mind that any action must be done guaranteeing the sustainability of the planet.

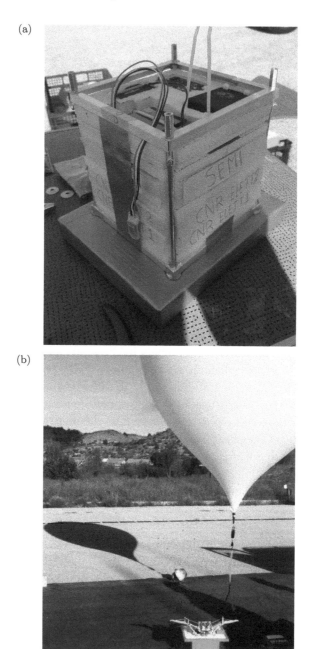

Fig. 8.1 Project Labonstrato: (a) box for hosting experiments, (b) Leonardo flying machine ready to fly.

In this book several points regarding the Leonardo da Vinci characteristics as complex systems engineer have been outlined. And examples of the model of his controlled machines have been presented.

Some time ago it was said that the XXI century was expected to be the age of Complexity Science, indeed just five hundred years ago Leonardo da Vinci left a great message on this. Writing this book we were able to uncover how Leonardo was the first Engineer of Complex Systems, investigating the Science of Complexity.

Acknowledgements

The authors would like to thank Riciclandia (Multitranciati srl) of Arezzo (Italy) and the Centro Internazionale di Studi sul Barocco di Siracusa (director Prof. Lucia Trigilia) where the machines are exhibited today.

Appendix A

Appendix I - Biographic Notes on Leonardo da Vinci

Leonardo da Vinci was painter, architect and scientist (Vinci, Florence, April 15th 1452 - Cloux castle od. Los-Lucè Amboise May 2nd 1519). Illegitimate son of the notary ser Piero, di Vinci, since 1469 settled in Florence where in 1472 he was already registered with the Company of Painters. According to Vasari, in 1476, Leonardo da Vinci was a pupil of Andrea del Verrocchio. Between 1482–1483, he was in Milan at the court of Ludovico il Moro, sent as *musico* (expert of musical instruments) by Lorenzo de Medici. It was a very fruitful period for Leonardo da Vinci, both for designing of works and for study that was interrupted by the defeat of Ludovico il Moro in 1500. He left Milan with his mathematician friend, L. Pacioli and the pupil Salaí and went to Venice, stopping along the journey to Mantua, at the court of Isabella d'Este, where he was welcomed with great honor and request for paintings. In 1501, he left Venice to return to Florence where he led a life not organized, dedicated to painting but more often gave "opra forte ad la geometria, impacientissimo al pennello" (very careful to geometry, impatient with the brush). He had already received commissions from the King of France Louis XII. From May 1502 to May 1503, he was far from Florence almost always in contact with duke Valentino (Cesare Borgia) and in close contact with Louis XII. A safe-conduct of Valentino declares Leonardo "Architecto et Ingegnero Generale" (General architect and engineer). Various notes highlight in this period several trips to Urbino, Rimini, Cesena, Cesenatico, and Pesaro where he studied harbors, rivers, canals, problems of hydraulics and fortifications. Back in Florence, he still dealt with military matters and canalizations, paintings and began to study the flight of birds. Disappointed at the unfortunate outcome of the great mural painting *Battaglia di Anghiari*, with the frustration of his engineering projects, and the incomprehension of Florentine artists and patrons of his research activity, Leonardo da Vinci returned again to Milan in 1505. From some drawings, it seems that Leonardo followed Louis XII in Brescia at the time of the battle of Agnadello (May 14th 1509), studying the hydrography of the region. In 1513, he was called to Rome by Guglielmo de Medici, but was excluded from the great works of that time. But Leonardo had not lost his contacts with France and in 1517 he took refuge with Francesco I, who gave him residence in the Cloux castle near Amboise. Leonardo had with him some pictures, an infinity of volumes of notes and though prevented by paralysis of the right hand, he worked with passion at the studies of anatomy, architecture and festive paraphernalia. On April 29th, 1519, he made a will: he died three days later. Most of the manuscripts he bequeathed to F. Melzi; many have come to us in arbitrary collections from by P. Leoni in the 16th century. The most important are at: Paris Institut de France Library: Ms. A; Ms. B; Ms. C (Codice di Luce et ombra); Ms. D; Ms. E; Ms. F; Ms. G; Ms. H; Ms. I; Ms. K (Codex Archintianus), Ms. L; Ms. M; Ms. Ashburnham 2038; Ms. Ashburnham 2037. Torino, Royal Library: Code on the flight of birds, Milano Biblioteca Ambrosiana: Code Atlantico, Sforzesco Castle Milan: Code Trivulziano. Holkham Hall, Count Leicester Library: Code Leicester. London Victoria and Albert Museum: Code Forster I, II, III. London British Museum: Code Arundel 263. Windsor Castle Library Royal: drawings and studies of anatomy.

Leonardo da Vinci wrote with his left hand to mirror — turning the letters and the words from right to left because he was left-handed (witness L. Pacioli 1498). The work of deciphering the manuscripts and drawings was completed starting from 1800 onwards, in particular by the da Vinci commission created in 1902. Despite the large number of autobiographical notes in the Vincian manuscripts, Leonardo's personality still eludes a clear definition. Particularly the relationship between art and science appeared already obscure to many contemporaries because while some complained that scientific concerns distanced Leonardo from the artistic realization others are surprised that Leonardo was "celebrato per pictura" (celebrated for paintings) but "obscuro in le altre parte, che sono a lui de grandissima virtute" (obscure in other aspects of his great virtue). Historiography restored the substantial unity in Leonardo's mind between scientific and artistic research.

Appendix B

Appendix II - Circuits and Microcontroller Codes

In this Appendix, the technical details related to the automation of the Leonardo machines described in Chapter 7 are proposed. In particular, the circuit schematics needed to interface the mechanical structure with the microcontroller are described, together with the codes necessary to program the control actions.

B.1 Bluetooth module

In several automated machines, we installed on the microcontroller a bluetooth module to allow for a remote control of the mechanics of the Leonardo designs. We selected a HC-05 bluetooth module, reported in Fig. B.1, which is produced by different manufacturers and it is easily available in online shops. The main advantages of this module are its efficiency, ability to provide Arduino microcontrollers bluetooth connectivity by using the internal supply of the microcontroller itself, and its low-cost. Moreover, it can be easily programmed by using open-source libraries available on the internet. We used the library **SoftwareSerial.h** and initialized the bluetooth module with the following code:

```
#include <SoftwareSerial.h>
#define  BT_RX 11          //  PIN receiving from bluetooth
#define  BT_TX 10          //  PIN transmitting to bluetooth

SoftwareSerial bluetooth(BT_RX, BT_TX);

void setupBT() {
    bluetooth.begin(9600);
}
```

In the codes reported as below, we adopted a similar syntax, which may vary if a different PIN for receiving and/or transmitting has been used.

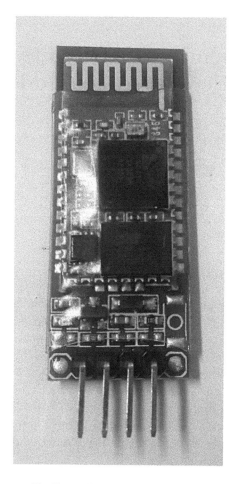

Fig. B.1 Bluetooth module HC-05.

B.2 Drop propagation

The drop propagation is actuated using a microcontroller which drives the DC motor. The code is:

```
#include <LiquidCrystal.h> /* managing LCD1602 */
#define MOTOR_PIN 5
```

```
#define BUTTON_PIN 3
#define POTENTIOMETER_PIN A0
#define BOOT_TIME 400

LiquidCrystal lcd(13, 12, 11, 10, 9, 8);
bool enabled = false;
float lastEnableValue = LOW;
unsigned long startTime = 0;
float vin = 0.0;
float giri = 0.00;

void setup() {
    pinMode(MOTOR_PIN, OUTPUT);
    pinMode(BUTTON_PIN, INPUT);
    lcd.begin(16, 2);
    lcd.setCursor(0, 0);
    lcd.print(" LA GOCCIA DI ");
    lcd.setCursor(0, 1);
    lcd.print(" LEONARDO v1.0 ");
    delay(3000);
    lcd.clear();
}

void loop() {
    vin = ((analogRead(POTENTIOMETER_PIN) * 5.0) / 1023.0) + 4.94;
    giri = ((vin * 5.2)-6.1);
    if (vin < 0.09) {
        vin = 0.00;
    }
    lcd.setCursor(0, 1);
    lcd.print("V=");
    lcd.print(vin);
    lcd.print(" RPM=");
    lcd.print(giri);
    delay(300);
    int enableValue = digitalRead(BUTTON_PIN);
    if (enableValue != lastEnableValue && enableValue == HIGH) {
        lcd.setCursor(0, 0);
        lcd.print(" Sistema ON ");
        if (!enabled) {
            startTime = millis();
        }
        enabled = !enabled;
    }
    lastEnableValue = enableValue;
    if (enabled) {
        if ((millis() - startTime) > BOOT_TIME) {
            int motorSpeed = map(analogRead(POTENTIOMETER_PIN), 0, 1023, 80, 255);
            analogWrite(MOTOR_PIN, motorSpeed);
        }
```

```
    else {
        analogWrite(MOTOR_PIN, 255);
    }
}
else {
    analogWrite(MOTOR_PIN, 0);
    lcd.setCursor(0, 0);
    lcd.print(" Sistema OFF ");
} delay(50);
}
```

The circuit schematic is reported in Fig. B.2, where a potentiometer as a voltage-divider is used to change the supply voltage.

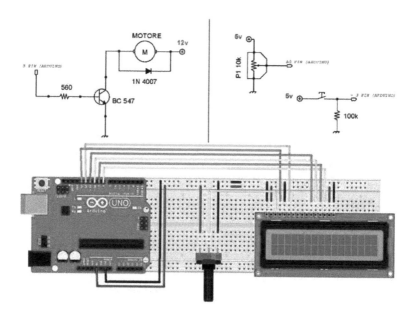

Fig. B.2 Schematics of the control circuit for the drop propagation.

B.3 Mechanical clock

The code regulating the mechanical clock provides the motor an ON or OFF signal depending on the response of the two sensors monitoring the position of the weight. Motor is in fact turned on when the weight reaches the bottom position and then turned off when the top sensor identifies the weight in the upper position. The code is reported as:

```
const int motorPin =  9; // the number of the motor pin
bool activated = false;
void setup()
{
  Serial.begin(9600);
  // initialize the motor pin as an output:
  pinMode(motorPin, OUTPUT);
}

void loop()
{
  double bottomSensor = analogRead(A5);
  if (bottomSensor<950)
  {
    if(activated == false)
    {
      digitalWrite(motorPin, HIGH);
      activated = true;
      Serial.println("moton ON");
    }
  }
  if(activated == true)
  {
    double upperSensor = analogRead(A4);
    if (upperSensor<950)
    {
      digitalWrite(motorPin, LOW);
      activated = false;
      Serial.println("motor OFF");
    }
  }
}
```

The schematic of the circuit devoted to control the motor and to acquire sensory information is as in Fig. B.3, where the IRF520N transistor used to drive the DC motor is seen, as well as the two optical sensors TCRT5000 and the relay which disables the motor.

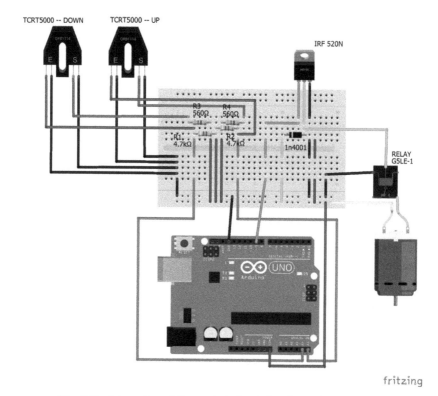

Fig. B.3 Schematics of the control circuit for the mechanical clock.

B.4 Lever crane

The lever crane is based on a servomotor controlled by the bluetooth module allowing a smartphone to give the direction of motion to the lever. Moreover, two buttons located at the extreme positions of the lever avoid the overrun of the lever implementing an autonomous cycle of lifting and lowering of the crane.

The code used to program the microcontroller is:

```
#include <SoftwareSerial.h>
#include <Servo.h>
#define pinb1 7
#define pinb2 8
#define  BT_RX 11
```

```
#define   BT_TX  10
#define   pinservo  9

Servo servo;
int i=1;
int j;

int servoposition;
int servopos;
int new1;
SoftwareSerial bluetooth(BT_RX, BT_TX);

void setup() {
    int pos=0;
    Serial.begin(9600);       // start serial communication at 9600bps
    bluetooth.begin(9600);
    Serial.println("--- Ports ready ---");
    servo.attach(pinservo);
}

void loop() {
    if (bluetooth.available()){
        String value = bluetooth.readString();
        servoposition = value.toInt();
        if (value.toInt() == 1) {
            while (digitalRead(7)!=HIGH && i<160) {
                i++;
                Serial.print(i);
                servo.write(i);
                delay(20);
                if (bluetooth.available()) {
                    value = bluetooth.readString();
                    Serial.println(value);
                    if (value.toInt()==3) {
                        Serial.println("YYY");
                        break;
                    }
                }
            }
            if(digitalRead(7)==HIGH || i==160) {
                servo.write(i);
            }
        }
```

```
if (value.toInt() == 0){
    while (digitalRead(8)!=HIGH && i>0) {
        i--;
        Serial.print(i);
        servo.write(i);
        delay(20);
        if (bluetooth.available()) {
            value = bluetooth.readString();
            Serial.println(value);
            if (value.toInt()==3) {
                Serial.println("YYY");
                break;
            }
        }
    }
    if(digitalRead(8)==HIGH || i==0) {
        servo.write(i);
    }
}
if (value.toInt() == 2){
    j=0;
    while(j==0) {
        while (digitalRead(7)!=HIGH && i<160) {
            i++;
            Serial.print(i);
            servo.write(i);
            delay(20);
            if (bluetooth.available()) {
                value = bluetooth.readString();
                Serial.println(value);
                if (value.toInt()==3) {
                    Serial.println("YYY");
                    j=1;
                    break;
                }
            }
        }
        Serial.print(i);
        if(digitalRead(7)==HIGH || i==160) {
            while (digitalRead(8)!=HIGH && i>0) {
                i--;
                Serial.print(i);
                servo.write(i);
                delay(20);
```

```
                        if (bluetooth.available()) {
                            value = bluetooth.readString();
                            Serial.println(value);
                            if (value.toInt()==3) {
                                Serial.println("YYY");
                                j=1;
                                break;
                            }
                        }
                    }
                }
            }
        }
    }
}
```

The circuit schematic is seen in Fig. B.4.

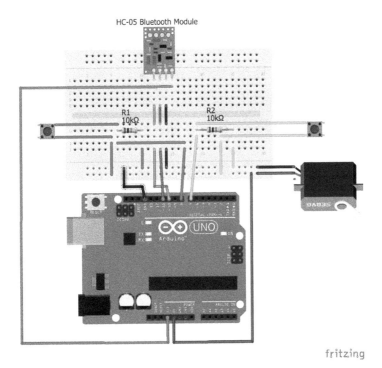

Fig. B.4 Schematics of the control circuit for the lever crane.

B.5 Mechanical drum

The mechanical drum was implemented as a remote controlled cart. The bluetooth module, the Arduino board and the batteries are located onboard the vehicle, therefore the model is completely autonomous. The code allows forward and backward actions, as well as turning of the wheels. The code adopted for the microcontroller is:

```
#include <Servo.h>
#include <SoftwareSerial.h>
#define ANGLE 35   // Define the Steering Angle
#define TIME 500
Servo servoMain; // Define our Servo
char ser;

const int controlPin1 = 2; // connected to pin 7 on the H-bridge
const int controlPin2 = 3; // connected to pin 2 on the H-bridge
const int enablePin=11;
const int MotorSpeed=255;

void setup(){
  // initialize the inputs and outputs
    pinMode(controlPin1, OUTPUT);
    pinMode(controlPin2, OUTPUT);
    digitalWrite(controlPin1, LOW);
    digitalWrite(controlPin2, LOW);
    pinMode(enablePin, OUTPUT);
    digitalWrite(enablePin, LOW);

    servoMain.attach(6); // servo on digital pin 11
    servoMain.write(40);

  //inizialize serial
    Serial.begin(9600);    // Open the Comunication
}

void loop(){
  // change the direction the motor spins by talking to the control pins
  // on the H-Bridge with make the pin 2 and 3 LOW and Motorspeed alternativily
    if(Serial.available()){
        ser = Serial.read();
        if(ser=='w' || ser == 'W'){
            servoMain.write(40); //set the initial position;
            digitalWrite(controlPin1,HIGH);
```

```
            digitalWrite(controlPin2,LOW);
            analogWrite(enablePin,MotorSpeed);
            delay(TIME);
            analogWrite(enablePin,0);
        }
        if(ser=='s' || ser == 'S'){
            servoMain.write(40); //set the initial position;
            digitalWrite(controlPin1,LOW);
            digitalWrite(controlPin2,HIGH);
            analogWrite(enablePin,MotorSpeed);
            delay(TIME);
            analogWrite(enablePin,0);
        }
        if (ser=='d' || ser == 'D'){
            servoMain.write(40 - ANGLE);   // Turn Right
            digitalWrite(controlPin1, HIGH);
            digitalWrite(controlPin2, LOW);
            analogWrite(enablePin,MotorSpeed);
            delay(TIME);
            analogWrite(enablePin,0);
            servoMain.write(40); //set the initial position;
        }
        if(ser== 'a' || ser == 'A'){
            servoMain.write(40 + ANGLE);   // Turn Left
            digitalWrite(controlPin1, HIGH);
            digitalWrite(controlPin2, LOW);
            analogWrite(enablePin,MotorSpeed);
            delay(TIME);
            analogWrite(enablePin,0);
            servoMain.write(40); //set the initial position;
        }
    }
}
```

The circuit needed to interface the microcontroller with the cart is shown in Fig. B.5, where note the adoption of a half-H bridge driver used to change the direction of motion from backward to forward, and viceversa.

B.6 Self-supporting bridge

The self-supporting bridge is used as a testbench for dynamic vibrational analysis. The vibration is induced on the bridge by using an electromagnet consisting of a solenoid with ... windings over a cylindrical support with diameter of 1cm, using 7m of isolated copper string (0.14mm diameter,

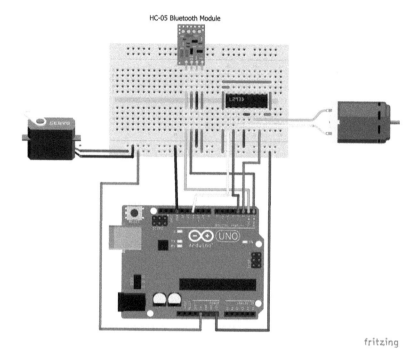

Fig. B.5 Schematics of the control circuit for mechanical drum.

0.015mm^2 area of the section) providing a resistance of 8Ω. Building a solenoid is an easy and cheap task and allows the control of the magnetic force either in voltage or in current. The permanent magnet coupled with the solenoid is a sintered Neodymium (NdFeB) magnet with a nickel plating (Ni-Cu-Ni) of class N45 providing a magnetic force of 2.5kg. The magnet is located on a fixed support in order to maintain its relative position with respect to the solenoid. The interplay between the magnet and the solenoid depends on the current, therefore a mutual repulsive action is generated by opposite magnetic fluxes, otherwise a mutual attractive action is due to coincident magnetic fluxes. The vibration is induced on the structure alternating the direction of the current through the solenoid according to the polarity of the square-wave control signal supplied by a standard function generator through which the frequency can be varied.

The power is supplied to the solenoids through an amplifier based on TDA7297. The circuit, whose schematic is seen in Fig. B.6(a), is a two-channel audio amplifier that is able to provide a power up to 15W per

Appendix II - Circuits and Microcontroller Codes

channel to an impedance of 8Ω, despite the fact that the control signals used have the main spectral density at a range of frequency below that for which the amplifier has been designed. We choose to adopt this scheme, whose implementation is seen in Fig. B.6(b), since it is easy, reliable and low cost. The gain of the amplifier can be set to act on a single potentiometer.

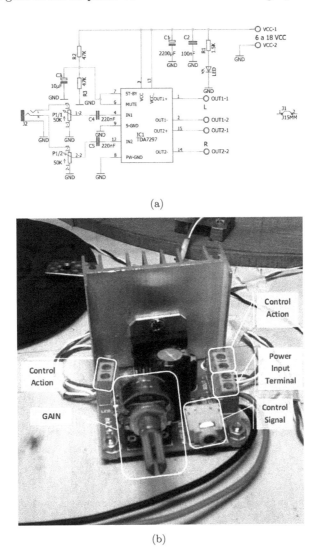

Fig. B.6 Control circuit for self-supporting bridge: (a) schematic of the amplifier, (b) picture of the implemented circuit.

B.7 Transformation of motion

The code used to control the model for transforming motion is here reported:

```
#include <SoftwareSerial.h>

int rxPin=3;
int txPin=2;
SoftwareSerial bluetooth(rxPin,txPin);
const int motorPin = 9;
int switchState = 0;
String message="";

void setup() {
  // put your setup code here, to run once:
    pinMode(motorPin, OUTPUT);
    Serial.begin(9600);
    bluetooth.begin(9600);
}

void loop() {
  // put your main code here, to run repeatedly:
    while (bluetooth.available()) {
    message="";
    message+=char(bluetooth.read());
    }
    if (!bluetooth.available()) {
       //switchState=digitalRead(switchPin);
       if (message!="") {
           if (message=="H") {
               digitalWrite(motorPin,HIGH);
               Serial.println(message);
               message="";
               delay(20);
           }
           else if (message=="L") {
               digitalWrite(motorPin,LOW);
               Serial.println(message);
```

Appendix II - Circuits and Microcontroller Codes

```
            message="";
            delay(20);
        }
    } else { Serial.println("0");}
  }
}
```

The circuit, as in Fig. B.7, contains an IRF520N transistor to provide enough current to the motor, the bluetooth module is included to allow the remote control.

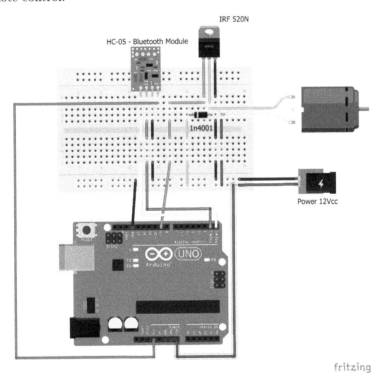

Fig. B.7 Schematics of the control circuit for transforming motion.

Bibliography

Ackerman, J. S. (1978). Leonardo's eye, *Journal of the Warburg and Courtauld Institutes*, pp. 108–146.

Argan, G. C. (1966). Enciclopedia universale dell'arte, *Florencia: Sansoni*.

Baratta, M. (1912). Leonardo da vinci e la cartografia. prolusione al corso di geografia letta nella r. università di pavia il dì 16 febbraio 1911.

Barenghi, C. F., Skrbek, L., and Sreenivasan, K. R. (2014). Introduction to quantum turbulence, *Proceedings of the National Academy of Sciences* **111**, Supplement 1, pp. 4647–4652.

Bhagat, R. K., Jha, N., Linden, P. F., and Wilson, D. I. (2018). On the origin of the circular hydraulic jump in a thin liquid film, *Journal of Fluid Mechanics* **851**.

Bucolo, M., Buscarino, A., Famoso, C., Fortuna, L., and Frasca, M. (2019). Control of imperfect dynamical systems, *Nonlinear Dynamics*, pp. 1–11.

Buscarino, A., Fortuna, L., and Frasca, M. (2017). *Essentials of Nonlinear Circuit Dynamics with MATLAB® and Laboratory Experiments* (CRC Press).

De Viguerie, L., Walter, P., Laval, E., Mottin, B., and Solé, V. A. (2010). Revealing the sfumato technique of leonardo da vinci by x-ray fluorescence spectroscopy, *Angewandte Chemie International Edition* **49**, 35, pp. 6125–6128.

Finsen, H. (2005). *Matisse: a second life* (Happy Books).

Firpo, L. (1971). *Leonardo: architetto e urbanista* (UTET).

Fortuna, L., Famoso, C., Buscarino, A., and Frasca, M. (2017). *Control of imperfect nonlinear electromechanical large scale systems: from dynamics to hardware implementation*, Vol. 91 (World Scientific).

Fuller, R. B. (2001). *Buckminster Fuller: anthology for the new millennium* (Macmillan).

Heydenreich, L. H. (1952). Leonardo da vinci, architect of francis i, *The Burlington Magazine* **94**, 595, pp. 277–285.

Kemp, M. (2007). *Leonardo da Vinci: the marvellous works of nature and man* (Oxford University Press).

Lupacchini, R. and Angelini, A. (2014). *The art of science: from perspective drawing to quantum randomness* (Springer).

Macagno, E. (1982). *La meccanica dei fluidi nei codici di Madrid di Leonardo da Vinci*.

Macagno, E. (1985). Leonardo's methodology in his fluid mechanical investigations, in *Proceedings International Symposium on Refined Modelling of Turbulence. Paper K*, Vol. 3.

Macagno, E. (1991). Some remarkable experiments of leonardo da vinci, *La Houille Blanche*, 6, pp. 463–471.

Maffei, L. and Fiorentini, A. (2008). *Arte e cervello* (Zanichelli).

Mainzer, K. (2007). *Thinking in complexity: The computational dynamics of matter, mind, and mankind* (Springer Science & Business Media).

Milano, M. L. (2019). Time machines, http://leonardo3.net/en/l3-works/machines/1426-time-machine.html.

Monaghan, J. J. and Kajtar, J. B. (2014). Leonardo da vincis turbulent tank in two dimensions, *European Journal of Mechanics-B/Fluids* **44**, pp. 1–9.

Moon, F. C. (2007). *The machines of Leonardo Da Vinci and Franz Reuleaux: kinematics of machines from the renaissance to the 20th Century*, Vol. 2 (Springer Science & Business Media).

Pevsner, J. (2002). Leonardo da vinci's contributions to neuroscience, *TRENDS in Neurosciences* **25**, 4, pp. 217–220.

Svaetichin, G. and Jonasson, R. (1956). A technique for oscillographic recording of spectral response curves. *Acta physiologica Scandinavica. Supplementum* **39**, 134, p. 3.

Taddei, M. (2007). I robot di leonardo, *Milano: Leonardo3*.

Tonomura, A. (1998). *The quantum world unveiled by electron waves* (World Scientific).

Vasari, G. et al. (1890). *Trattato della pittura di Leonardo da Vinci condotto sul cod. vaticano urbinate 1270* (Unione cooperativa editrice).

Vinci, M. L. (2019). http://www.museoleonardiano.it/ita/.

Weyl, H. (1950). *The theory of groups and quantum mechanics* (Courier Corporation).

Wolf, M. and Stoodley, C. J. (2008). *Proust and the squid: The story and science of the reading brain* (Harper Perennial New York).

Young, G. (2012). *Manual specialization and the developing brain* (Elsevier).

Zeki, S. and Nash, J. (1999). *Inner vision: An exploration of art and the brain: Oxford university press* (Oxford).

Lightning Source UK Ltd.
Milton Keynes UK
UKHW020608030220
358062UK00003B/6